BUSINESS IN GREAT WATERS

They that go down to the sea in ships: and
occupy their business in great waters:
These men see the works of the Lord: and
his wonders in the deep.

(Psalm 107)

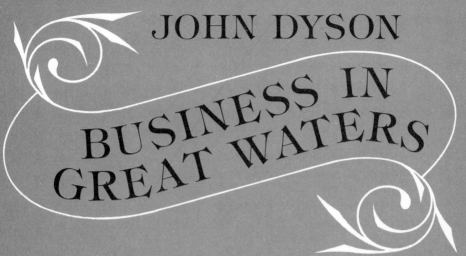

JOHN DYSON

BUSINESS IN GREAT WATERS

The Story of British Fishermen

ANGUS & ROBERTSON · PUBLISHERS

ANGUS & ROBERTSON . PUBLISHERS
London . Sydney . Melbourne . Singapore . Manila

First published by Angus & Robertson (U.K.) Ltd 1977

Copyright © John Dyson 1977

The eight drawings on pages 133, 151, 177, 181, 201, 221, 278, 297 © Iain Rae

ISBN 0 207 95706 1

Set in Monotype Old Style No. 2

Previous books by John Dyson

Fiction

The Prime Minister's Boat Is Missing

Non-fiction

Yachting The New Zealand Way
Save The Village Pond
The Magnificent Continent (part author)

Children's Books

The Pond Book
Fun With Kites (with Kate Dyson)

For my son, Jack

Move over John, let Jack sit down
(Fishermen's saying)

CONTENTS

Preface   9
Foreword   11

I

HEALTH T' MEN AN' DEATH T' FISH

*How the nature of the sea and its fish have evolved a unique type of man –
the fisherman*

Shipmates of Old Nick   15
Pork of the Sea, Beef of the Sea   25

II

LAUNCH OUT INTO THE DEEP

*Despite her maritime traditions Britain's fishermen were for centuries
backward and irresolute – until the North Sea boom ushered in a golden
age of silver harvests*

A Subtle Contrivance   35
Sterling Herring   43
Scotch Cure   49
Peggoty's Port   61
Cod Bangers   66
Silver Pits   73
The Great Emporium   82

III

IRON MEN, WOODEN SHIPS

*Great fleets of sail fished the Dogger Bank and quartered the North Sea
from the 1830s to the end of the century: it was the golden age – romantic,
simple, but cruel, violent, monotonous*

What Cheer, Ho?   91
Reef and Tops'l Breeze   104
Sentenced to The Dogger   113
Devil's Work   119
The Bethel Ships   128
Over for the Lord   145
The Real Price of Fish   159

IV

THE SEA HATH FISH ENOUGH

Before the First World War fishermen in coastal villages made a bare living, but these comparatively golden years were the last of fish in plenty

Fifie, Skaffie and Zulu	171
Petticoat Government	184
The Coble Coast	196
Spat, Sprat and Bawley	206
Bogs and Hoggies	214
Mackerel Metropolis	220
Meat, Money and Light	229
Nickeys and Nobbies	234
The Best Sauce is Appetite	238

V

NET PROFIT

As steam took over from sail the sea could no longer produce fish enough; fishermen became Arctic scavengers chasing fewer fish with larger ships and 200-mile fishing limits imposed a radical new outlook

Sign of Zodiac	245
Cement, Iron-rust and God's Mercy	257
Not Fish Enough	269
Pipe Stalkies	275
Dover Patrol	286
World War to Cod War	292
Appendixes	313
Bibliography	324
Picture Credits	329
Index	330

PREFACE

A century and a half ago a North Sea boom was being discussed in the same tones of awe and excitement with which it is spoken of today. Then it was not oil that had been discovered, but herrings, soles, plaice, and other kinds of fish. On the strength of it Britain's fishing industry became the envy and admiration of the world. But little is known of those who toiled on the grey waves – men hardy and adventurous beyond compare, yet conservative and emotional to a degree. In the span of a lifetime their way of life has virtually disappeared. Of that large fishing community that lived on Britain's shores during the boom years of the industry, from the end of the Napoleonic Wars to the First World War, only memories and faded postcards remain.

This is the story of how fisherfolk of the nineteenth century lived and worked, and I have attempted to put it in the context of what happened before and afterwards. It is largely a story in two parts, that of the trawlerman who spent – and still spends – long periods at sea and was a stranger to his home, and that of most other fishermen who lived outside the main ports and for whom fishing was not just a vocation but a way of life. There is also a marked difference between Scottish fishermen and their counterparts in England, but Wales has little in the way of a fishing tradition. The story concentrates on herring and white fishery. Whaling and commercial salmon fishing are outside the scope of this work.

Earliest records do not go much further back than Elizabethan times and statistics do not begin until the end of the last century, so much of the story is based on first-hand accounts of observers. Fishermen were not noted for keeping logs or journals, unlike other mariners, and the researcher has to tread a wary path between a true account of an amazing incident and a good yarn with a grain of truth. No book can hope to be fully comprehensive, but I have opted for the human side at the expense, perhaps, of a strictly comprehensive economic review of the fishing industry.

The many books and papers which were consulted in the course of research are listed on page 324, but I must single out for mention the wealth of material contained in Edgar March's three books about fishing vessels, Peter Anson's books about Scottish fisherfolk, and John Leather's account of smacksmen in Essex and Suffolk. I am also grateful to A. N. Meldrum and *The Countryman* for the account of life in a Scottish fishing port on page 194. When quoting from other books and newspapers, particularly those written before the turn of the century, I have taken the liberty of shortening and making small alterations for readability, but in no case has the sense been changed.

The search for material and photographs led to many dusty cellars

9

and slushy fish quays, but no source was more profitable than that of the Royal National Mission to Deep Sea Fishermen. Its secretary, Mr Jack Lewis, and his kindly staff, helped me to search through the mission archives and it is through their generosity that I am able to make use of so much interesting and valuable material, including sketches reproduced from back numbers of the mission's magazine *Toilers of the Deep*.

The Herring Industry Board was most helpful, and I would particularly like to thank Mr Robert Mure and the Board's chairman, Dr W. J. Lyon Dean, who told me about the building of the first Zulu fishing boat by his grandfather. The White Fish Authority's Hull office provided a great deal of assistance, and I would like to thank Skipper Ken Knox of Kingfisher Charts, and Mr R. C. Markham of the U.K. Trawlers Mutual Insurance Company Ltd, also at Hull. Mr Jonathan Watson Hall, of Thomas Hamling and Company Limited at Hull, helped me to talk to trawler owners. I was especially grateful to spend a day with Skipper Jack Ellis, now retired, who told me of his own experiences in Arctic waters, and to read the first draft of the book written about him by Dick Capstick. Mr and Mrs John Buchan of Peterhead provided a lot of information about the Scottish fleets, and for those interested in the subject I cannot do better than recommend a visit to the Scottish Fisheries Museum on the quay at Anstruther, Fife, which is one of the few places where the spirit of the age of fleets of black-sailed drifters does live on; I am indebted to Mr Gordon Clarkson, Mr Garth Sterne, and Miss Mary Murray for their hospitality and assistance during my long visit. Also, I would like to thank the Fishmongers' Company for the use of its fine library; Mr Gerald Gardner of the Shellfish Association of Great Britain; Mr Alan Viner of the National Maritime Museum at Greenwich; Mr Charles Lewis, curator of the Great Yarmouth Maritime Museum, who gave me an opportunity to visit the steam drifter *Lydia Eva* that has been restored by the Maritime Trust and is open to visitors in summer; Mr Edward Paget-Tomlinson, of the Maritime Museum, Hull; the staff of the library at the Ministry of Agriculture and Fisheries in Whitehall; the staff of the British Library; Mr Parry Watson and Mr Peter Allard of Great Yarmouth; and Lt. Col. John Willcocks. The book owes a great deal to the beautiful pictures drawn by Mr Iain Rae of Aberdeen, who is a professional ship-modeller, and Mr K. C. Lockwood of Southend.

John Dyson
London
February, 1976

FOREWORD

During my last six years in the Navy, I spent a good deal of time in the North Sea and the North Atlantic comfortably ensconced on the bridge of my cruiser flagship. We often passed the fishing trawlers, which seemed very small to us, rolling and pitching horribly with their crews working unconcernedly about the decks and pausing to give us a cheery wave.

My admiration for fishermen with their cheerful acceptance of a hard and dangerous life dates from this period. Since my retirement I have worked for ten years with the Fishermen's Mission and have learned a great deal more about fishermen and their families, who need the same courageous outlook on life as do their men at sea. And through my association with the National Maritime Museum I have learnt something of the different craft they use around our coasts, and of the history of their development.

But had I had a copy of *Business in Great Waters* ten years ago, I should have been spared a lot of miscellaneous reading, and should have been much better informed. This excellent book tells all there is to know about the history of the fishing industry, of its boats, of its crews, and of the harsh fluctuations in their fortunes which in an ever changing world have never stayed secure for long.

I hope it will reach a wide public and would wish too that it will be read by the politicians and officials in whose hands, at this critical period in the industry's history, the future of the fishermen's livelihood lies.

Admiral Sir Charles Madden Bt., G.C.B., D.L.
Chairman of the Trustees of the National Maritime Museum.
Chairman of the Royal National Mission to Deep Sea Fishermen.

I

HEALTH T' MEN AN' DEATH T' FISH

The sea's our field of harvest,
Its scaly tribes our gain,
We'll reap the teeming waters
As at home they reap the plain.

Shipmates of Old Nick.

Fisherfolk believed that when old fishermen finally cast off their moorings and sailed on the ebb for a rendezvous with Old Nick they returned – as seagulls. From the behaviour of seagulls haunting pier-heads, fish docks and the long-shore, the possibility does not seem at all unlikely. You only have to look at a line of seagulls hunched against the breeze to be reminded of tough old men with linseed-oiled sou'-westers worn at every angle. They have the same wary attitude towards strangers. Like fishermen, seagulls work in crowds but are fierce individualists. In the seagull's wind-ruffled strut you can see the old

Old fishermen believed gulls were dead shipmates reincarnated.

fisherman's indifference to a bit of wet; in the cocky tilt of its head as it floats buoyantly over thundering breakers the fisherman's contempt for the sea's rage is clearly recognisable. When a seagull steals another's morsel of fish guts the furious shrieks of rage echo the old fisherman's talent for language – 'That's Dinger telling Buck and Ponzy to lay off, he always was an irascible old sod.'

If we could tap the shyly gruff conversations of gulls wheeling along the shore and following fishing boats out to sea we would hear sadness. The old fisherman swooping down to perch on his own rooftop would most likely find it gone: the higgledy-piggledy tarry-smelling cottages that served generations of fishing families have been swept away or turned into curio shops. Gone, too, are the ramshackle sheds made of wooden boats cut in half and turned upside down, the grassy strips where thousands of drift-nets were laid out to dry, and the smoggy swirl of pungent smoke draughting downward from the louvres of quayside smokehouses. Within living memory fleets of red, brown, black and golden sails dotted the horizons by the thousand: they have all gone, and so have the herring that were netted by the hundred million.

The fishermen of Britain have had few champions and even fewer glorifiers. As if stick-drawn on damp sand, the patterns of a community way of life that persisted for many centuries have been swept clean by the spring-tide of progress. Centralised, industrialised, modernised, fishing continues to thrive and fishermen are as skilful and resourceful as ever, but what has been lost is the unique life-style that functioned around the fisherman and his boat; sadly there are not even many relics of that age. The steam-locomotive era survives in a number of railway lines restored by enthusiasts; narrow boats and canals are reminders of how folk lived on the inland waterways; watermills, smallholdings, beam engines and Thames barges typify other distinctive ways of life. Of fisherfolk very little remains – except the seagulls.

Until about the 1920s fisherfolk were almost a race apart. Despite its small area Britain is remarkable for its diversity of regional characteristics in her people. Cornishman and Yorkshireman, Scot and southerner, share nationality and a common heritage yet until their regional differences became blurred by transportation and education they met as foreigners, had different beliefs and customs, and lived as separately as people on far-flung islands.

Fishermen were unique because their way of life allowed them to transcend these differences. A man was first a fisherman, second a seaman, and third a Yorkshireman, Scot, Cornishman, or whatever. In a herring fleet gathered at a port such as Whitby you could identify the big and ponderous Lowestoft man, like the Yorkshireman slow of gesture and phlegmatic, the nimble and chattery Highlander, the reserved and determined Cornishman. But these regional differences were small compared with the qualities and idiosyncrasies that singled him out from others because his livelihood was the catching of fish.

A fisherman was distinctive in any crowd, regardless of his regional origin. He was not instantly recognisable merely by virtue of a sou'-wester like a wrinkled helmet, a patterned jersey large enough to fit a barrel, seaboot stockings into each of which a boy might creep, or boots large enough to contain fair-sized halibut. It was in his gait, deliberate, as if the ground might suddenly develop a list, his arms held out from the side as if carrying buckets and those enormous blubbery, insulated fingers half-clenched as if ready to pick up a rope. The black and juicy pipe burning hotly on a short stem appeared to serve a double purpose, for he could feel the warmth of its glow on his face. Invariably it was a face that betrayed a stolid indifference to bodily comfort, a flinty determination, a certain sense of wisdom, and a fatalism tinged, perhaps, with sadness. It was not a face you would see in a committee: a fisherman might work in a fleet for convenience but his whole life was competitive. It was the face of a man who, when a 'drop o' water' falls aboard like a cartload of bricks, does no more than blink and spit.

If is easy to be romantic about the figure cut by the fisherman, particularly the shaggy-whiskered individual of early memory whose quaint attire identified him as certainly as any uniform. The patterns of his square dark-blue jersey or guernsey, knitted by wife or sweetheart, had particular meanings: rings and furrows on the waistbands or neck denoted how many children and grand-children he had, anchors denoted hope, zig-zag lines were lightning or symbolic of the ups and downs of married life. On Sundays the patched fall-front trousers of heavy serge gave way to a neat suit of navy-blue pilot-cloth, the coat with a velvet collar and the trousers slightly bell-bottomed, perhaps (as at Great Yarmouth) with a two-inch notch in the seam of each cuff so it fell better over tight and squeaky boots.

Fishing for a living is a vigorous, thrusting and economically risky occupation which has few virtues and the single crowning disadvantage that it is hazardous in the extreme. Many men in ordinary life have had to face real danger, particularly those who served in the armed forces, but it is seldom that an individual knows real fear on more than two or three occasions in a lifetime. A fisherman works constantly in conditions of appalling hazard. He stares death in the face in a way that most of us can barely conceive and knows that before long he will once again be in a situation of mortal fear. It is perhaps this that gives the fisherman a look about him that seems to say that he considers other people to be only half alive.

More noted for language than for conversation, a fisherman has always been taciturn by nature and it is a hard job getting him to admit the hazards of his calling. Danger is a simple reality, just as the fact that 'water's hellishly wet'. A fishermen is a hunter, and the simple truth of fishing is that unless he ventures from the safety of a harbour and casts his net or lines in the sea he won't catch any fish. In fact, fishing is probably the only branch of sea-going in which recklessness is not only permissible, but is an essential ingredient of success. It is recklessness of the sort that compels the good fisherman to shoot his trawl for another haul with a force nine gale breaking around his ears in the hope that 'a bit of breeze' will stir the fish into his net. It is the kind of recklessness that sends a fisherman to sea during a gale so that he will be ready and in position to commence fishing as soon as the wind moderates. There is another kind of recklessness, too: the kind that makes a skipper stop fishing a patch where he is getting average hauls and steam perhaps a couple of hundred miles to a new area to look for even bigger catches. Now he risks not his life but his job and reputation, but by dedicated fishermen these are counted as the same thing.

The daily routine of shooting and hauling nets from small vessels tossing about in the open sea is a display of supreme technical seamanship accomplished as a matter of course. Fishermen have always been a cut above other mariners in the sense that seamanship is only a means to an end, a technique that enables them to fulfil a more important and quite different function, which is to catch fish. The Navy goes to sea to patrol, and if needs be to fight. Merchant ships journey from one port to another. Only the fisherman has to grapple in such a personal way with the sea itself.

The many and various skills necessary in fishing are reflected in a typical fishing vessel, such as a trawler. In order to fill its holds with fish a trawler is a great deal more than a seaworthy floating platform. First, it is a submarine hunter. Second, it is a cargo ship, unique in finding its cargo from the sea, at sea, going out with empty holds and full bunkers, returning with full holds and empty bunkers. A trawler spends more time than any other vessel stopped at sea, so it has the sea-keeping qualities of a lightship. To drag a heavy trawl along the sea-bed it must be as powerful and as manoeuvrable as a tug, yet to rush its cargo to market it is also a high-speed vessel, comparable with the Cunard liner Q.E.II in terms of length/speed ratio. It is a lifeboat, required not only to stay afloat in dreadful sea conditions but to continue working in weather that drives all other types of vessel into shelter. Above all, it must be reliable, for a trawler spends some three hundred days of the year at sea.

Master mariner, tow master, lifeboatman, and above all a hunter,

---

1. *Whitby fisherman, c.1880, wears sou'wester of linseed-oiled cotton, and traditional close-knitted gansey of thick dark-blue wool; patterns had traditional meanings and also served to help identify drowned fisherman if his body was recovered.*

FMS. 248.

the fisherman himself is skilled in many maritime trades. The average fishing boat, however, is an untidy looking hulk, streaked with rust and spotted with fish scales. In port it is a jumble of wires, ropes, boxes and old gumboots. Only at sea, working for its living, is it a beautiful object of function and design, but few people see it there. Prodigious feats of seamanship accomplished at sea go unremarked, but when the fisherman steams home and kisses the quay with a bit of a thump the pier-head professors are likely to dip their yachting caps in sad recognition of the fisherman's obvious shortcomings.

The fundamental paradox of the fisherman's character has always been the necessary dash of recklessness combined with an absolutely iron-bound conservatism. A fisherman will risk anything except try something new. This contrasts oddly with the nimble and articulate instincts of the hunter: a man who lives by his wits could be expected to welcome new ideas that might well make his job easier or his own life safer. Until the electronic age made a technician of him (in addition to his traditional skills) the fisherman was the last to accept any refinement of a technique that had served perfectly well. It could be said that the fisherman's virtue of persistence scored his brain, or that continuous hardship dulled his vision.

Characteristic of this was fishermen's automatic refusal to admit any merit in such things as steam capstans, steam propulsion, motors for small boats, lifejackets, decked boats (in Scotland), the use of bags of oil for calming the sea, and radio sets. Fishermen have characteristically hated and scorned all protective measures. 'We're used to being drowned' was their attitude, and they hated more than death itself the chance of being ridiculed. There was a fixed idea that to wear a lifejacket was a sign of cowardice, yet a fisherman's idea of bodily comfort was to swathe himself with layers of clothes, weighing in total some twenty-five pounds, and when he took what was called the fisherman's walk (three steps and overboard) his enormous thigh-length boots with heavy nailed and iron-heeled soles filled with water and he was carried straight down with little chance of swimming, even if he knew how. On the one hand the old-time fisherman was insensible to fear; on the other hand he feared nothing more than the jeer of another fisherman.

A fisherman's station in life was invariably working class: from earliest times only the very poor would take on the hard work and discomfort. Even in the 'good old days' from about 1840 until 1914, when Britain's fishing industry was at the peak of its prosperity, the reality of hardship, danger and drudgery in a fisherman's life is scarcely imaginable. Yet fishermen were infinitely better off than the average farm workers and factory employees of the day, among whom social injustice and malnutrition was rife.

A remarkable feature of small coastal fishing communities (as distinct from the big trawler ports) was that whole families were involved in the work, particularly in Scotland, where a man without a mother or a wife to bait his hooks and sell his fish was at a great disadvantage. Fisherfolk were clannish, proud, and although they lived in poor communities they knew a great deal about the world because they saw comparatively alot of it and worked alongside fishermen from other nations. Their houses might have been crude and ramshackle but they were clean and there was never any excuse to starve. Unlike work in the factories, their jobs were wholesome activities carried out in the sun and the wind. It might have been dangerous, but on the whole it was a healthy way of life.

From an economic point of view fishing was always unsteady. A fisherman was at the mercy of the market, and first he had to get his catch to the salesman before it spoiled. Transport was expensive, markets were often depressed. In a community fishermen would help each other in launching boats and loading gear; they often fished in company and at sea mutual aid was an absolutely instinctive reaction. But when it came to selling their fish it was every man for himself, and fishermen would never band together in some form of co-operative. They were interested only in the prices they could get when their fish were landed; even if a fish bought from them for a shilling made a guinea at Billingsgate, the idea of clubbing together and sending their fish to Billingsgate themselves was never entertained.

The fishing family was the epitome of self reliance. Even children played their part in equipping the boat for fishing by collecting bait. Men mended their own nets, dipping them in boiled cutch then laying them out to dry. They cured fish for their own use and for sale, tarred their own vessel, did their own repairs. Despite storms that left scores of widows and hundreds of orphans unprovided for, and accidents that occurred with the regularity and inevitability of changes in the weather, it was rare that fishermen took the trouble to insure themselves; the fishing community looked after its own.

A fisherman at least had the satisfaction of an independent way of life; he might have been held in a rut by economic circumstances but it was nowhere near as deep as those in which miners and mill-workers were trapped for the best parts of their lives. Fishing was one of the few fields in which a boy could work his way up from the poor-house to own a fleet of vessels, and many did. Even today, one of the attractions that draws men to trawling is the possibility of making a great deal of money as a skipper.

With the coming of trawling fleets in the North Sea, in the middle of the last century, the story of fisherfolk in Britain becomes divided: trawlermen led a very different life to the fisherman with his own boat in a small coastal community. The trawlerman spent most of his time at sea and was a stranger to his own home. He did not live in a cottage as close to his boat as possible, but in a terraced house in a big town,

2. *Whitby fishermen exchanging a light for their clay pipes on the stern of a smack, c.1880; note the massive tiller and rudder post.*

like any industrial worker. As the sailing smacks were superseded by expensive steamers he did not even have the satisfaction of working for himself, or knowing that he, too, could one day become an owner. In trawling, fishermen were reduced to the status of hired hands, and became inclined towards waywardness and irresponsibility which was not a feature of the closely knit traditional fisherfolk communities. In port they still tend to live flamboyantly, making the most of a precious few hours ashore, and for this they are sometimes unfairly condemned.

From about the First World War until the 1960s, the trawlerman dominated the fishing scene while other forms of fishing seriously declined due to sewage pollution, wrecks, over-fishing, and changing tastes in favour of trawler-caught cod. Now, however, the pendulum is swinging the other way and Britain's inshore fishermen are doing well while trawling is becoming increasingly depressed and Hull owners report a loss of ninety pounds for every day that a vessel spends at sea. A young man can again (with government aid) become skipper of his own vessel, but competition is tougher than ever and to be successful he must have what it takes to be a fisherman – courage, persistence, a 'nose' for fish, supreme seamanship, and that essential streak of recklessness.

Pork of the Sea, Beef of the Sea.

Seventy per cent of the world's surface is covered by sea – 331,000,000 cubic miles of water. If one cubic mile of sea water was loaded into railway trucks it would make a train 834,000 miles long which at forty miles an hour would take two years and five months to go by. . . . In one cubic mile of sea water there is 120,000,000 tons of common salt, 2,200 tons of iodine, 900 tons of copper, 25 tons of gold. . . . If all the salt was extracted from the sea it would cover the land to a depth of 500 feet. . . . The world's largest creatures and biggest plants live in the sea, with four-fifths of known animal species (excluding insects) and at least 20,000 species of fish.

The sea produces about the same amount of vegetable matter every year as the land, but most of this grows over the shallower continental shelves where the water is less than 100 fathoms (600 feet) in depth. The shelf which surrounds the British Isles extends more than a hundred miles south-westwards into the Atlantic. Near the coast an acre of sea grows five and a half tons of vegetable matter in a year, compared with one and a half tons of hay cut from a good field of the same size.

The 'grass' of the sea is plankton, myriads of tiny plants (and animals) which drift about in clouds but are invisible to the naked eye except as a green tint. If you tow a fine net through sea water the slushy green stuff, collected, like atomised spinach, is a mass of tiny, single-celled plants that multiply very rapidly, particularly in spring when the sea blooms like a meadow. Plankton survives on oxygen which is absorbed into the water through the agitation of the waves on the surface, and salts washed into the sea in huge amounts by rivers. The animal plankton feeds on the plant plankton and the English Channel produces nearly 4,000 tons per square mile every year.

25

Floating about on the surface near the light these specks of drifting life are the basic foods of bigger creatures, such as sprats and herrings, which in turn are preyed upon by larger fish such as cod. The rain of waste products from plankton on the surface manures the sea-bed which provides food for other creatures such as shellfish, crustaceans, worms and crabs, which are also eaten by fish. Britain is fortunate in being surrounded by very shallow seas in which plankton flourishes; in agricultural terms this gives her the advantages of a rich dairy farm and market garden over a barren piece of moorland. Also, the 'climate' is ideal because the circulation of the ocean north of the equator is anti-clockwise, with the result that warm water is constantly streaming in, melting the ice as far north as Spitsbergen. Contrast this with the opposite side of the Atlantic: in Labrador, which is on the same latitude as southern England, the sea is frozen for half the year.

The North Sea has potential as a vast reservoir of food comparable with the fertile prairies of North America. It fulfilled this function until recent times, when it became fished out and was turned into the marine equivalent of a dustbowl. Its average depth is only 360 feet, one-thirtieth that of the Atlantic, and in many places, such as the Dogger Bank, the spire of a village church placed on the sea-bed would be visible above the surface. The Dogger Bank itself, which has been so important in the development of British fisheries, is an immense submarine bank, remnant of a recently sunken and still sinking part of Europe, about as large as Yorkshire and Lancashire. It is riddled with hills and valleys, like the countryside of Devon, and fishermen came to know its bumps and hollows, and those of other North Sea banks, as well as they knew the street-plans of their home towns.

Fish are thought to have lived in much the same forms as we know them today for much longer than Man and other animal species. Their streamlined and torpedo-like shape (except in the case of flat-fish) is Nature's answer to the fact that water is 800 times heavier than air. Its weight increases with depth so at 3,000 fathoms the pressure is three and three-quarter tons per square inch, yet because the pressure acts equally and in all directions at once even the most fragile fish can survive. Every fish has an air bladder which acts like the trimming tank of a submarine and by the automatic adjustment of gases a fish can select any depth. Fish are incredibly voracious and always on the prowl but can go months without a meal. They use up less energy than a land animal because they do not have to support themselves and therefore need less rest. They are cold-blooded, their body temperature is about two degrees higher than the surrounding water, and they are short-sighted with blurred, monocular vision. Their transparent scales are protective, acting like an overlapping coat of mail, and the slime that covers them is antiseptic, to help ward off fungus and bacteria, while also reducing friction between the fish and the water. From the rings in the scales, like the rings of a tree, you can tell the age of a fish.

There are two main types of fish. *Pelagic* fish swim in huge shoals at certain times of year, often just beneath the surface, where they feed on plankton; they are 'fat' fish, storing oil in their bodies, and include the herring, mackerel, sprats and pilchards. *Demersal* fish, or 'white' fish, live on or near the bottom, feeding on small creatures including other fish; they store oil in their livers and include all the important trawled and line-caught fish such as cod, haddock, and flat-fish.

Apart from the fact that cod eat herring, the two 'worlds' of fish are as distinct from one another as pigs and cattle. Over the centuries quite different methods of fishing have evolved, and this in turn has had an influence on the development of different fishing communities: fishermen of East Anglia and Scotland are historically linked with herrings, and Cornishmen with other pelagic fish such as mackerel and pilchards; the major trawling ports such as Hull and Grimsby owe their importance to demersal fish, first flat-fish such as soles trawled in near waters, then vast supplies of cod and haddock found in Arctic waters. Pelagic fish, in the main, were trapped by the gills in the meshes of long curtains of net hanging just below the surface; demersal fish were caught by hook and line, or scooped up by funnel-shaped trawl nets dragged along the sea-bed by long ropes or wires.

The herring's presence in, or absence from, the North Sea has affected the destiny of the world. The Hanseatic League founded its fortune on herrings in the Middle Ages. Holland became a world maritime power on the strength of its herrings, most of which were caught in or near British waters. Until recent times the herring was the most plentiful of fish: in 1913 over a million tons were taken from the North Sea, half by England and Scotland. Congregating in certain areas at different times of year to spawn they formed immense shoals and in 1887 the naturalist T. E. Huxley estimated there was one fish per cubic foot: in a shoal one mile square and three fathoms deep there would be 500,000,000 herring – and some shoals were eight miles long.

Beautifully adapted for living in upper water layers, with a blue-green back matching the depths when seen from above, and a silver belly that blends with the bright gloss of the surrounding waters when seen from below, the herring is the most important food fish. There are more calories in a pound of herrings than in equivalent amounts of meat or eggs. Three herrings provide all the protein and fat a man requires to survive a working day, and he will leave the table feeling that he has eaten well. At the present time the National Health Service spends millions of pounds a year on prescriptions for vitamins, iron preparations and 'tonics' yet most of the ingredients – including calcium, iron, iodine, Vitamin B_{12} which is the 'anti-anaemia' vitamin, and the rest of the B complex, and Vitamins A and D – are all found in quantity in herrings.

From the Middle Ages until the middle of the last century herrings were a staple food for rich and poor, but to eat them required a strong palate for the fatty flesh of this 'pork of the sea' deterioriated rapidly. A powerful mustard sauce was the usual way of disguising the unpleasant tang of putrefaction. As soon as herrings were landed on the beach they were normally sprinkled with rough salt and stirred around with a shovel, which delayed the rotting process. When soaked in water for at least twenty-four hours, then boiled, salted herrings were palatable although they tasted like dish-cloths filled with fine bones; nevertheless, they were good enough for the archers at Agincourt who went into battle on breakfasts of salted herrings.

When it was two days by coach from Dover to London, and five days from Hull, opportunity to eat fresh herring inland was small. Wherever fishing was a matter of subsistence, in Scotland and most of the coastal communities of England, fish were the best eating. Herring fresh out of the sea and into the pot, or laid spluttering on the grid-iron, was simply delicious. Split open, lay face to face, dip in oatmeal and fry, was a typical cottage herring recipe from Scotland.

The other standby was stockfish – cod lightly salted and dried on stocks in the sun and bitter-dry winds of Newfoundland. It had the taste and melt-in-the-mouth qualities of an old sou'wester, the fibres were extremely indigestible and almost impossible to chew, and its consistency was so timber-like that fishermen fashioned it into intricate caskets and other gifts but few examples of this art remain because supplies inevitably ran short on long voyages and the handiwork went into the pot. Dessication removed the fishy odours and changed the flavours so it had a bacony and oily taste, and when ground into a powder made a tolerable sauce. It remained in good condition a long time and supplemented the salt-meat diet of Elizabethan sea-dogs on their voyages of conquest and exploration. At this time England was better known for its stockfish and saltfish than for its roast beef.

Cod, the 'beef' of the sea, is the commonest and most important demersal fish and is abundant everywhere north of the British Isles. There are nineteen common species, including the cod itself and haddock, coal-fish (coley) or saithe, pollack, whiting and blue whiting. Practically all parts of the cod have a use. In ancient times the tongue, fresh or salted, was a great delicacy; gills were preserved for bait; the roe was another delicacy; dried bones fuelled the fire; the head of a big

---

3. *Old boots, herring shovels, rope—a quayside shop in Whitby, c.1880. The ridges knitted into the arm of the shop-keeper's gansey traditionally indicated his number of children.*

4. *Cod—the "beef" of the sea. A fine specimen caught by the stern trawler* Sir Fred Parkes.

cod, fried whole, would feed several children; the liver oil, first used medically in 1782, was also a substitute for whale oil for fuelling lamps and dressing leather. In modern times fish heads are used for glue, fish meal, paint and plastics; skins for glue and fish-leather; the scales for the manufacture of fire extinguishants and artificial pearls; the gall bladders for manufacturing bile acids and the livers as a source of hormones, vitamins and pharmaceuticals.

Until the early nineteenth century cod was eaten only as stockfish, cured in Newfoundland, Iceland or northern Scotland. The smaller haddocks, which existed in great numbers in the North Sea, were thrown overboard from fishing vessels until it was found they could be smoked. Supplies of fresh fish to London increased with the development of well smacks which kept fish alive after they had been caught by hook and line, but 'live' (that is, only recently dead) cod and other fish such as plaice and turbot were very expensive and sold only in places like Bath and London's West End. In Queen Anne's time fish could be sent directly to Billingsgate from Great Yarmouth overnight, but it took twenty changes of horses to do it. The cost of transporting fish from Lyme Regis and Tor Bay was equal to the cost of the fish it-self – fish that in summer might already have been lying in the hold of a boat becalmed in sultry weather for a couple of days before it was landed to spend another twenty-four hours in a closed cart dashing as best it could along roads that had been better in Roman times.

Odours of putrefaction from 'fresh' fish stalls in the streets were said to have been the cause of the Great Plague of 1664, despite the fact that a few years earlier James I had granted the Fishmongers' Company powers to inspect any fish 'to view wither the same be wholesome for Man's Body and fit for sale, or no.' In 1771 Tobias Smollett said that when it arrived in London fish 'would turn a Dutch-man's stomach, even if his nose was not saluted in every alley with the sweet flavour of fresh mackerel selling retail.'

The ordinary person had little chance of obtaining fresh fish unless he lived on the coast itself, or on the route of a fish-wife hawking her husband's catch around the outlying parishes; but even if you were rich freshness was a commodity money could seldom buy. When Henry IV married Joan of Navarre in 1403 three of the six courses at their wedding banquet were fish. Among such delicacies as *perys in syrippe* and *creme of almaunds* were *lampreys powderyd* (spiced), *rostyd* and *in paste; trouteg; samoun; gurnade; plays; breme; purpayis* (porpoise) and *crabbe.* Henry VIII ordered daily for his table fresh quarter of seal, porpoise, herrings (salt and fresh), cod, ling, stockfish, halibut, turbot and flounder. But how many of these fine-sounding delicacies would be palatable it is impossible to say; probably very few.

The art of properly salting fish, and hanging it over a hot and smoky fire to cure it, had been known to the ancient Egyptians who depicted

30

it on murals, and the Phoenicians had a flourishing trade in cured fish. The ancient Greeks detested salt fish and fished only in a small way, but no Roman banquet was complete without its fish course (including British oysters) and fish-salting became a huge industry employing thousands of slaves. In Europe the art was not discovered until the fifteenth century and in Britain its development was hindered by injudicious taxes on salt. Smuggling salt became an industry almost as great as that of fishing when the first salt taxes were levied by Elizabeth I, and these grew more and more onerous until, by 1782, taxes were 4,000 per cent on the salt's real value. During the Napoleonic Wars salt in Glasgow cost 5d. a pound, or £45 a ton; when salt tax was abolished in 1825 (only because it was handicapping the chemical industry) the price fell from £32 to £1 in a decade and by 1880 was only twelve shillings a ton.

Fisherfolk and villagers near the coast smoked, salted and dried fish for their own consumption and many preferred it to fresh fish. Up to the middle of the last century every house or cottage would have its few fish split and opened out and hanging on the wall to dry out of reach of children and dogs. In the West Country conger eels were split and opened out then stitched end to end to form long lines of drying fish. In Highland villages travellers encountered disagreeable smells from lines of haddock and whiting hung to dry on lines along the sides of houses. In the north of Scotland, and the Orkneys and Shetlands, cod and ling and other white fish were salted for three days then dried on pebble beaches, being placed in heaps at night, and finally were stored between layers of fresh oats and straw.

With the expansion of the railways and the use of ice for preserving fish at sea the picture changed radically. In the space of a few years fishing was transformed from a fragmented, badly organised cottage industry into an immense mechanised, capital-oriented, systematic hunting operation involving steamships, railways, and chains of food stores. The days when every grocer had a barrel of salted herrings by the door and a pile of stockfish in a dark recess of his shop began to disappear now that fresh fish was easily and cheaply available, even in the great industrial cities of the North and Midlands. Between 1851 and 1911 the population of Britain doubled and the availability of cheap fish filled the gap left by the over-fishing of oysters.

Meanwhile, in the middle of the last century, a new product began to be hawked about the streets of London – fried fish, sometimes sold with a slice of bread or a baked potato. At first fried fish was sold only to the very poor: costermongers bought up left-overs from fishmongers' slabs, fried the fish as a means of preserving it, then wheeled it round the streets, often for several days, before it was sold at a penny a piece. Sales were best in the gin-drinking neighbourhoods – 'For people hasn't their smell so correct there.'

When potato chips were introduced from France in the 1870s it also became economic for sellers of fried fish to supply fresh, trawled cod. Hot pie shops began to make a speciality of fried fish and chips. Soon the fish and chip shop had spread to every neighbourhood, particularly in the sprawling industrial areas. The fish was not of prime quality but any shortcomings in appearance were hidden beneath layers of batter. Steam trawlers filled the need. Just before the First World War the 25,000 fish and chip shops in Britain consumed a quarter of the country's total catch of 800,000 tons of white fish.

Frozen fish had long been known in Russia and Canada where it was displayed in shops in chunks of ice and cut up with a saw when required. Apart from one or two unsuccessful experiments between the wars it did not become established in Britain until the early 1950s, which happened to coincide with the arrival of commercial television and modern marketing methods. It is still argued today whether television advertising ensured the success of frozen fish-fingers, or whether massive promotions by frozen-food companies made television advertising. The result was a completely new trend in Britain's fish-eating habits.

Now more than half Britain's consumption of fish is cod, and more than half the cod is frozen. With catches of cod declining due to international quotas and the Icelandic restrictions, we face the prospect of having to do without cod almost entirely. The British housewife may think she faces disaster, but the worst that can happen is a return to the kind of fish diet enjoyed by the Victorians in the days of sail. About 1850 only 1,800 tons of fresh cod was landed at Billingsgate compared with exactly the same weight of sprats and 10,000 tons of mackerel.

To be thrown entirely on our own resources for fishing might be inconvenient and it would certainly be difficult, but it would not be disastrous. The sea off Britain's shores is so fertile that with modern farming methods and new techniques it can be a substantial source of food. In 1915 the government estimated that an acre of the best mussel ground yielded 10,000 lbs. of mussel meat with a value of three million calories a year which was worth £250, but an acre of good pasture produced only 100 lbs. of beef with 120,000 calories which was worth £17 10s. Even if Britons do not take to frozen fingers of mussel meat, or mussels and chips, mussels could provide food for fish reared or 'farmed' in enclosed waters such as the Scottish lochs. There are many other kinds of fish, such as blue whiting and sprats, which exist in large quantities but are unpopular for no good reason. Tastes will undoubtedly have to change, but it is unlikely that we will ever again have to resort to the 'fisherman's sauce' of mustard and vinegar to disguise the taste of rotting fish.

II

LAUNCH OUT INTO THE DEEP

Ye gentlemen of England who live at home at ease,
Oh! Little do you think upon the dangers of the seas.

A Subtle Contrivance.

As instruments of Man's survival fish-hooks and nets pre-date the plough. Fishing for the pot is at least as old as hunting, and throughout history until modern times it has remained a primitive and almost instinctive adventure. Until the Norman Conquest ordinary people in England could fish in tidal waters where they pleased. In common law the dominion and jurisdiction of the sea was vested in the sovereign, but the right to navigate on the sea, and catch fish in it, belonged to the subjects. With no pollution, no urban sprawl, the many estuaries around the British coast so teemed with fish that it was hardly necessary to go afloat to catch them: the greatest numbers were caught in fixed traps built of stones, stakes, brushwood and nets on low-tide sands and mudbanks, and across the mouths of rivers and creeks.

The general principle of the diversity of fish traps that evolved around the coast was that fish were allowed to pass inward, or up-river, with the flood tide, but were trapped when the tide ebbed. On broad sands such as Morecambe Bay, Minehead, and the Menai Straits, very large weirs were built in intricate shapes so that the fish were led into puzzle gardens at the centre of which was a trap in the form of a deep pool; this had to be emptied immediately the tide receded to prevent the gulls having a feast. The common form of fish trap or weir was v-shaped, with each arm up to four hundred yards long. The barriers accumulated great quantities of silt, the stakes becoming thick with mussels and seaweed, and lasted for scores of years.

Although it was subjected to increasing control and regulation this method of fishing persisted for centuries, not only on the coast but in rivers for catching salmon. The various ingenious devices, known in legal terms as 'fixed instruments' or 'fixed engines' were known in Cumberland as *garths*, in Morecambe Bay as *baulks*, in Cardigan Bay as *goryds*, in Rye Bay as *keddles* or *kettle nets*, and elsewhere as *keppers*, *gurgites*, *raw fag nets*, *guorts* and *net weeles*. Fishermen had been able to exercise their ingenuity in trapping fish unrestricted by laws or taxes until the crown realized the potential for revenue in this widespread

35

and necessary occupation. Even as late as 1720 Daniel Defoe saw a permanent fish trap on the River Dart near Totnes where young fish in a foot or so of water left at low tide were chased into a hoop net by a trained dog.

At about the time of the Norman Conquest the freedom of the individual to fish any tidal waters was withdrawn. On the Continent the ruling princes licensed fishermen who had to pay fees, on the grounds that fishing inevitably led to disputes from which subjects had to be protected. Somehow the idea crossed the Channel and the Domesday Book listed seventy-six fisheries in tidal waters controlled by the Crown; most were at Benfleet, Clacton, Tollesbury, Tilbury and other creeks in Essex. When Henry II ascended the throne in 1154 fisheries were extensively controlled. On nearly every stretch of tidal water in England fishing rights had been granted by the Crown to particular individuals, estates and town corporations. How the ordinary man was prevented from poaching is not known. The Crown's privilege was ended by the Magna Carta in 1215, leaving a legacy of contention that continued for centuries.

In some cases the rights had been divided, certain people having the rights to oysters and others to swimming fish; one might have an exclusive claim to red fish (salmon) but white fish (herring, sprat, etc) belonged to another. Legal terminology was unclear: for centuries lawyers argued the precise definitions of these ancient rights, attempting to draw distinctions between 'several fisheries', 'free fisheries', 'common fisheries' and 'common of fishery' until the House of Lords decided they all meant the same thing – 'the rights to the profits of the soil over which the water flowed in which fish were taken.' In other words, while shellfish could be owned, swimming fish were a common resource and like any wild animal did not belong to an individual until they were taken out of the sea, when they became the property of their captor.

After Magna Carta the proliferation of fish traps in estuaries and rivers became a serious menace to navigation. Small ports were completely closed to ships because of fish traps built across their entrance channels. The Magna Carta itself demanded the demolition of all fish weirs everywhere except on the coast itself, but the problem remained a persistent headache for successive monarchs. The many laws enacted had only short-term effects and it was not until 1605 when James I prohibited the construction of any fish trap within five miles of the mouth of any haven that control became more effective.

From Elizabethan times salmon rivers were protected by conservancy commissioners, close seasons and minimum size regulations, but the control of fishing in tidal waters was in the hands of an odd assortment of courts of admiralty that exercised jurisdiction in a variety of ways. In the Thames and Medway, for example, fishing was controlled by the Corporation of London; the towns of Ipswich and Colchester regulated fishing on the Colne and Blackwater, and some lords of manors on the coast had similar courts, such as the Manor of Brancaster, Norfolk. Each court laid down its own regulations regarding mesh sizes, close times, and the use of particular types of trap. Boat fishermen who sailed off-shore to catch fish were not affected.

Concern for the protection of small fish and spawn, particularly of salmon, dates from earliest times. In the Thames the capture of immature fish was illegal as early as 1320, when sixteen nets belonging to Barking men were taken out of the river and burned on the orders of the mayor and aldermen of the City of London. The fishermen continued to set nets in defiance of the City rules so inspectors were sent out to enforce the regulations. In 1349 fifteen nets containing three bushels of fish 'which, by reason of their smallness, could be of no use to anyone' were confiscated at Barking and Greenwich.

Open rebellion occurred in the lower reaches of the Thames in 1406, when an inspector named Alexander Bones seized sixteen nets belonging to men of Barking, Erith, Pratt's Ferry and Woolwich. His plan was to lay them before the mayor for inspection but the fishermen rushed ashore, rang the church bells until a crowd of two thousand had assembled, and 'with bows, arrows, bucklers and clubs did pursue him in divers boats' to Barking. In danger of being mobbed, the inspector handed the nets to some constables but the angry fishermen took them back by force. Ultimately the ring-leaders were found guilty but a large committee of lords, including the Archbishop of Canterbury, decided to be lenient and allowed the same nets to be used until Easter on the understanding that they would then be surrendered and burned in the presence of the mayor.

The fishermen's angry reaction was due partly to their anxiety over a new and 'subtily contrived instrument' which was threatening their traditional methods of fishing on the river – the 'wondyrchoun'. A simple form of trawl, it had been in use for about seven years when Edward III was petitioned in 1376 about the decline of fish it was causing. The king was told that it was made in the manner of a large oyster dredge, on which was attached a net so close-meshed that no fish could escape. 'The great and long iron of the wondyrchoun runs so heavily and hardly over the ground when fishing that it destroys the flowers of the land below the water, and also the spat of oysters, mussels and other fish upon which the great fish are accustomed to be fed and nourished. By which instrument in many places the fishermen take such quantity of small fish that they know not what to do with them; and they feed and fat their pigs with them, to the great damage of the Commons of the Realm and the destruction of the fisheries.'

A commission appointed by the king met in Colchester the following year and reported that the net called the wondyrchoun was three fathoms long and ten feet wide, and had a beam ten feet long at the end of which were two frames shaped like 'colerakes'. A leaded rope weighted with big stones was fixed on the lower part of the net, which had meshes the length and breadth of two thumbs. The commission advised that the net ought to be used only in deep water and not in estuaries; no legislation was enacted but most private fisheries banned its use. However this outcry against the trawl proved to be only the first shots of a centuries-long siege aimed at having the trawl banned. In the 1620s Charles I was forced 'to consider the great destruction made of fish by a net or engine now called the trawle' and the argument continued through Victorian times with parliamentary commissions appointed to investigate the issue. The efficiency of the trawl always won the day, and in essentially the same form as the wondyrchoun it has been the prime tool of British fishing up to the present day.

Primitive forms of boat such as dug-out canoes were used in very early times in calm waters, but the extent of fishing from boats in open waters, and the methods used, do not come into historical focus until Elizabeth I arbitrarily ordered her subjects to eat fish so that the industry would prosper and in turn encourage ship-building and provide a reservoir of skilled seamen for the navy. On the east and south coasts small boats had fished at sea at least since the Viking invasions. Where the Danish marauders settled, communities inherited the traditions and skills of superb small-boat seamanship. Even today the fishing 'cobles' used on the coasts of Northumberland and Yorkshire betray in their proud bow and sweeping sheer the characteristics of the Viking longboat. It was certainly the Viking spirit and instinct of navigation that sent English fishing boats as far north as Iceland in the thirteenth century. During the reign of Edward I, (1272–1307), the Suffolk port of Dunwich, (now swallowed up by coastal erosion), sent twenty ships a year to Iceland and others sailed regularly from Bristol, Kings Lynn, Scarborough, Boston and Cromer.

Ketch-rigged English 'doggers', with a carrying capacity of about thirty tons and crews of five to ten men, sailed for northern waters in spring with supplies of salt and goods for trading, and spent the summer catching cod on hand-lines. The fish were salted and dried on board or ashore and brought home towards the end of the year then stored for sale during Lent. Some vessels did not fish but merely traded for fish caught by local people, and in 1420 a Danish edict determined the value of Icelandic cod: three fish for a pair of woman's shoes,

5 and 6. *Trawling and drifting in the 17th century from primitive boats off the coast of France: equipment improved and vessels became bigger but the methods of fishing did not significantly alter until after the Second World War.*

Fig. 2.

fifteen fish for a firkin of English honey. However, Denmark was intent on stopping this trade by English 'trespassers' and in 1432 England agreed that her fishermen would not fish north of a line drawn between Bergen in Norway, the Shetlands, the Faroe Islands and Iceland without a licence. As conditions became increasingly restrictive, and crews of English doggers faced a situation that had an almost exact parallel (though for different reasons) five centuries later when British trawlermen were engaged in three cod wars with Iceland, a new dimension appeared suddenly in the far west.

In 1497 John Cabot and his son Sebastian sailed from Bristol with a crew of eighteen local men to seek the spices and gold of Cathay. What they found were not seven fabled cities, but foggy inlets and dripping pine forests of Newfoundland and Labrador where the sea swarmed with cod – 'so many that this Kingdom would have no further need of Iceland.' Within a few years there was an annual 'cod rush' of ships sailing for Newfoundland's cheerless coast from west country and south coast ports such as Weymouth, Poole, Southampton, Exeter, Bideford, Barnstaple and Plymouth as well as London and some east coast ports. For four centuries, until Arctic trawling got under way seriously in the 1890s, Newfoundland dominated the distant-water fishing scene.

Ships fitted out over Christmas and sailed early in the year, the perils of a North Atlantic winter ahead of them. First to arrive, with nets, cordage, salt and other equipment, had the pick of the best harbours and landing places where stages of poles were built for drying the fish. All summer the men fished for cod with hand-lines and set lines. It was said you could hardly drop a weighted line overboard without touching the back of a cod. The fish were split and dried and lightly salted. Liver oil was produced in large open vats of decomposing livers, the golden oil being skimmed off as it rose to the surface.

The French were forced away from the coast by the hostility of English fishermen and found the immense Grand Banks where a man with a hand-line could catch four hundred cod in a day. With plentiful supplies of good-quality solar salt French fishermen gutted and split their cod and salted it in barrels on board, avoiding the necessity to land in Newfoundland and enabling ships to make two voyages a year.

In the Newfoundland cod fishery it was every man for himself. Early arrivals sabotaged the stages, small boats and liver vats left behind over the winter by other fishermen and late arrivals could waste a month looking for timber with which to repair their boats and build cabins. English fishermen set fire to hundreds of acres of forest, destroyed landing stages and other property, and spoiled more than forty small harbours by dumping ballast from their ships to make room for fish. There was bitter rivalry between itinerant fishermen and those who remained in permanent cabins over the winter. The settlers, known as 'planters', let their pigs and stock roam along beaches where cod was drying; fishermen burned their cabins and stole their food.

In autumn the summer's catch of dried and lightly salted cod – stockfish – was loaded aboard, equipment such as vats and small boats was hidden for the next season, and the ships sailed for Spain, Portugal, Italy and Greece where the fish was traded for spices, wine, velvets and other fine things which were taken home and sold: by this means England tapped the wealth of bullion that Spain had found in the New World.

The home fisheries were meanwhile at a low ebb. While men of bold and adventurous commercial spirit embarked on hazardous voyages to distant Newfoundland, fishermen at home seemed blind to the fleets of foreign vessels accumulating great wealth from herrings and other fish caught just off their own beaches. In 1580 an exasperated Elizabeth I banned the import of foreign-cured fish (exempting the Shetlands and Newfoundland) to stimulate English fishermen, but within seventeen years the law had to be repealed – 'the national subjects of this Realm not being able to furnish the tenth part of the same with salted fish of their own taking; the chief provisioning and victualling thereof with fish and herrings hath ever since been in the power and disposition of aliens and strangers, who thereby have enriched themselves.'

The dash of the Elizabethan fighting sailor was conspicuously absent from his neighbour the fisherman. While Dutch herring busses fished 2,000 strong off the British coast for half the year, exporting herring to twenty states and dukedoms (including England) and able to catch ten to twenty lasts[1] of herrings in each boat's nets which stretched for half a mile, English herring fishermen in a few score of ill-repaired boats fished only seven weeks with nets less than a quarter of a mile long and 'if we take seven lasts in a night it is a wonder.'

To generate demand for fish, for the benefit of her fishermen, Elizabeth I revived for purely political motives the religious idea that had disappeared with the Reformation – compulsory non-flesh-eating days. On 152 days a year, Fridays, Saturdays, ember days and Lent, and later on Wednesdays too, people were prohibited by royal decree from eating flesh of beast or fowl: this left no alternative but fish.

On a fish day a meal cost $1\frac{1}{2}$d. compared with 3d.; fish was cheap but hard to get, and when available it was seldom fresh. Few people could afford the menu offered at Lord Cecil's table on a fish day: first course of ling and salmon, or herrings, pike and plaice, or whiting, haddock and cod; second course of conger or lamprey or fish-roe and smelts; third course of tart.

Fish days were never popular, partly because they hinted too strongly of Papism. They proved impossible to enforce, and soon lapsed. In a scathing pamphlet Tobias Gentleman, a wealthy resident of Great

---

[1] See measurement table on page 314.

Yarmouth, estimated the total profit to the Dutch of exporting the so-called British herring at £4,500,000 – 'O slothful England and careless countrymen, look on these fellows that we call plump Hollanders, consider their activity and your own sloth. . . .'

Sterling Herring .

Herring was more than a staple food of catholic Europe. From the Middle Ages until the late eighteenth century herrings dominated commerce in a way that is comparable to the importance today of North Sea oil. Nations with a flourishing herring fishery became rich and powerful. It is one of the ironies of history that the biggest herring fisheries existed in British waters. Although she was a powerful maritime nation Britain was unable to protect her own waters, and it was the beginning of the nineteenth century before British fishermen finally learned the trick of catching and properly curing their own herrings on a large scale.

The earliest 'great' herring fishery of which records exist was on the south-west tip of what is now Sweden but was then part of Denmark. It reached considerable proportions by the end of the twelfth century and continued for three hundred years. Every September and October thousands of oared boats and scores of cargo ships voyaged to the area and huge populations of fishermen set up temporary camps along the shoreline. When fishing was already declining about 1500, because the shoals were mysteriously deserting the area, 7,515 boats took part, each with five men; the total manpower involved in catching, curing and transport was probably 100,000. Coastal land was divided into areas marked with boundary stones where representatives of various Baltic and north German states built their own store sheds, curing yards and even churches. Fishing was rigorously controlled: merchants of Lübeck brought their own executioner to impose justice. The short fishing season was so important that Pope Alexander III granted special dispensation to work on Sundays and Holy Days. Towns such as Lübeck and Bremen became immensely wealthy and combined in what was known as the Hanseatic League. Their merchants traded their salted herring and beer far and wide. In London their representatives had a warehouse where Cannon Street Station now stands and every year presented two barrels of herrings to the Lord Mayor. Their currency was so secure and reliable that other London merchants

43

often stipulated payment in pounds of the 'easterlings' which later became known as pounds sterling.

In England before the Reformation herring was the only cheap and readily available alternative to meat on the many Church fast and holy days. Some of the larger monastic establishments had their own fishermen, boats and storehouses. During the reign of Edward the Confessor, (1042–66), the Suffolk town of Beccles, then a small herring-fishing port on the shore of a wide and shallow estuary, paid a rent of 30,000 herrings a year to the Abbey of St Edmunds. This was just one of about 500 religious houses which in all had some 25,000 residents. In 1286, when an ox could be obtained for 4s. the abbey spent £25 on a stock of herrings for Lent.

Herrings were also used, together with stockfish, to supply castles and military establishments. During the Hundred Years War herrings were a basic food for the army. In 1429 there was a skirmish that became known as the Battle of Herrings: the ancient equivalent of the Royal Army Service Corps was taking five hundred cart-loads of herrings to the army of the Duke of Suffolk besieging Rouen when it was ambushed. The supply column formed a stockade of herring barrels and beat off the attack. The herrings reached the soldiers who went on to defeat the French and burn Joan of Arc at the stake.

The herring capital of England from the fifth century until the 1960s was Great Yarmouth. In the beginning the sandy peninsula on which the town now stands was only a sand-bank at the mouth of an estuary seven miles wide. The estuary covered what is now the Norfolk Broads but gradually it silted up and the sand-bank at its mouth became attached at its northern end to the mainland; the network of rivers and the saltings of Breydon Water behind the present town are all that remains of it. The sand-bank was permanently settled in 495 AD by a marauding Saxon war lord named Cerdic who landed the crews of his five ships. Every autumn the waters off that part of the coast of East Anglia seethed with immense shoals of herring, and fishermen of many nationalities, including those of English ports further south, made a habit of landing on the waste of tussocky sand dunes to dry nets and cure herrings. In 1208 King John formally recognised the rights of foreign fishermen to dry nets there without payment, and foreign merchants were allowed to buy fish. Another charter was granted in 1262 enabling the now thriving town to be enclosed on three sides by a wall and moat; the fourth side faced the sea. The denes became a sprawling depot for fishermen and merchants from Holland, France, Flanders and Norway, together with fishermen from the Cinque Ports of Sandwich, Dover, Hythe, Romney and Hastings (and later Winchelsea and Rye). The annual autumn gathering became one of the great free fairs of the Middle Ages.

The men of the Cinque Ports, who had traditional landing places and proprietary rights over certain parts of the denes were the most important and powerful body of mariners in England. Their position opposite France gave them every opportunity to harass the French, indulging in piracy and plunder in the name of the king. In return for acting as the maritime fighting arm of the king, in an age when there was no formally organised navy, the portsmen were able to enjoy important privileges. In 1260 Henry III granted them control of the herring fair. The king wanted the portsmen to keep the peace between native Englishmen and the great number of foreigners; the consequence, however, was almost total war between local men and the king's favourites from Kent and Sussex.

After preliminary skirmishes the Yarmouth men were conceded a half and equal share in the organisation and administration of the fair. But each time the Yarmouth men put to sea in their small boats they were harried by the larger and more powerful guard ships belonging to the Cinque Ports. Before long however the importance of the Cinque Ports declined; their towns suffered severely at the hands of the French, their rivers began to silt up. But their jurisdiction of the Yarmouth herring fair continued until the middle of the seventeenth century, although their visits ultimately became merely social occasions, with a certain amount of jostling for precedence at official functions.

For its help at the Battle of Sluys Great Yarmouth was granted the privilege of sharing its coat of arms with the king – three herrings with lions' heads. Then complaints reached the king about Yarmouth people forming price rings that worked against the herring fishermen and penalised consumers. In 1357 Edward III enacted the Statute of Herrings, which applied to any town in England where herring was landed but was framed particularly around Yarmouth to prevent local merchants making excessive profits as 'middle men'.

The irritation of the Cinque Ports bailiffs and the penalties of government price control were nothing, however, to the perennial problem of the state of Great Yarmouth's harbour. The great sea-arm was now reduced to a silt-laden river which formed shallow bars across its mouth. After heavy rain the river was liable to take an entirely new course, cutting through the sands then falling to leave a river mouth that was very wide, but so shallow you could almost wade it. Between 1347 and 1567 no fewer than seven new harbour entrances were dug through the accumulating silt by men with shovels. The first silted up again after twenty-six years and the second did not last long. In 1408 Henry IV granted one hundred pounds a year for five years towards the upkeep of the haven, which was draining the town of money, but two new entrances had to be dug. In 1548 the town sold the gold plate, bells and testaments of its church and raised £1816 19s. 7d.; the chief engineer of Dover was appointed to advise and another harbour was dug but the engineer died and the money ran out. Nine years later floods swept through the impoverished town, silting the harbour so badly that to regain the open sea ships had to be hauled over the denes by capstan. The seventh and existing harbour entrance was constructed between 1559 and 1567 by a Dutch engineer who brought 3,800 tons of stone from France to build solid piers that ran into the sea at right angles from the shore, creating a tidal scouring action that prevented silting. More than a thousand men, women and children worked on the task and a channel with ten feet of water at low tide was created. About 150 years later Daniel Defoe admired it as 'the finest quay in England, if not in Europe, not inferior even to that of Marseilles itself.'

The second great herring fishery of the North Sea was that of the Dutch. Probably in 1486, as the Scania herring fishery was beginning to decline, a Dutchman called Willem Beukels[1] made a simple discovery. Formerly herrings had merely been sprinkled with salt to delay decomposition long enough for fish to be landed then distributed by pack-horse or cart. It was a hit or miss method, and the herrings were usually tainted by the time they reached the consumer. What Beukels discovered was that herrings gutted immediately and laid carefully in good-quality barrels topped up with brine became pickled, travelled well, and lasted a very long time. The simple secret, which was to elude British fishermen until the early nineteenth century, made Holland a nation rich and powerful enough to shrug off the yoke of Spain, to explore and establish her own colonies, and to challenge the maritime power of Britain herself. Her fortunes built on nothing but the bones of herrings, Amsterdam became a great and prosperous city that rivalled London as a centre of world influence and trade.

Watching Dutch vessels lifting herrings out of the North Sea by the ton, it was little wonder that figures like Tobias Gentleman fumed at the 'slothfulness' of his own countrymen. In 1614 he wrote that Holland was equal only to Norfolk and Suffolk in size, and had to import everything from iron and timber for her ships to barley and beer for her people, but the wealth she earned from British herrings enabled her to war with Spain and despite the cost to grow rich and strong in fortified towns and beautiful buildings – 'and by reason of their industrious fisher trade not one of their people is idle, nor even seen to beg amongst them, except they be some of our own English nation.' It was less Holland's handsome profit than the insult to national pride that so irked Tobias Gentleman: 'We are daily scorned by these Hollanders for being so negligent of our profit and careless of our fishing, and they daily flout us that be the poor fishermen of England, to our faces at sea, calling to us and saying, "Ya English! Ya zall or oud scove dragien!" which in English is this, "You English! We will

---

[1] Spellings differ widely.

make you glad to wear our old shoes!" '

The Dutch themselves boasted that the herring fishery was a gold mine, that herrings kept Dutch trade going and Dutch trade set the world afloat. They exported herrings to most of the known world and even to Brazil. Their ships came home with oil, wine and wool from France and Spain, best salt from Portugal, velvets, spices and silks from the eastern Mediterranean, corn, wax, hemp and iron from the Baltic. And gold – the annual catch of 300,000 lasts was worth £10–£12 to the fisherman but when exported could fetch as much as £36 a last, leaving an overall profit to the merchant of at least £10. A little less than half was sold to England.

By 1601 Holland was reputed to own 20,000 sail, more than that of England, Scotland, France, Spain, Portugal, Italy, Denmark, Poland, Sweden and Russia combined. The herring fleet numbered about two thousand busses (pronounced bushes) – low, barge-like, cumbersome ships crewed by about fifteen men and able to carry about thirty-five lasts (sixty tons) of herrings. They had conspicuously high masts with square mainsails, running bowsprits with jibs and fore-staysails, and small topsails. Nets were laid in fleets more than half a mile long and were made of best hemp, or coarse Persian silk. A buss stayed at sea for long periods, making two or three separate voyages during the six-month season. Heavy to work, and slow and ponderous to sail, the busses sent their catches home by small, fast cargo boats called 'jagers': two or three jagers worked in the service of each buss, carrying salt and casks outwards and returning with pickled herrings. About five hundred Dutch boats traded every year to London with cod and ling, also caught in the North Sea, and a fleet of probably three thousand smaller vessels fished off Holland's own coast.

The herring fleet was organised as rigorously as any navy and the pickling was carried out under stringent and wise regulations to ensure quality. At first moor salt, obtained by burning briny turf from saltings near harbour entrances, was used. Later better salt was purchased from Portugal and it was carefully selected and inspected before being taken to sea. The salt was duty free, and during the season herring fishermen were privileged and could not be arrested for any debt unconnected with the catching of herring. Nearly half a million were employed, ashore and afloat – a fifth of the country's population.

The fleet assembled at Shetland and fished there from mid-summer day until July 25, when it moved to the eastern tip of Scotland just north of Aberdeen until September 14. The season ended with seventy days of fishing off the coast of Norfolk, during which the fleet was based at Great Yarmouth, the only town in Britain to benefit from Holland's North Sea gold mine. For 200 years Yarmouth serviced the Dutch herring boats in the same way that for the last ten years it has been a depot for the North Sea oil industry. During the herring fair ten thousand Dutchmen came ashore to buy corn and beans to take home, and bread, meat and other provisions for their vessels. Forty-three brewers were kept in business making beer for Dutch fishermen, so at least some of the English profited from their commercial rivals.

Scotch Cure.

The Dutch mastery of the North Sea herring fishery was fought in two ways simultaneously. With one hand the English and Scots tried to beat the 'plump Hollanders' at their own game by entering into direct competition, which was a royal flop. With the other hand they made aggressive gestures and deployed naval forces to restrict Dutch activities; this was a protracted business and led to three naval wars with Holland but ultimately, in the sense that the Dutch fishing effort was greatly diminished, it was successful.

Until the beginning of the seventeenth century England had laid no particular claim to territorial seas. While Elizabethan sea dogs asserted the English presence in the New World, at home the issues were rather different. Belief in the freedom of the seas was new, and people were uncertain of the issues. It was supported in a literal sense by Elizabeth I when Denmark tried to prevent the English from fishing near Iceland, and she instructed her ambassadors to declare that the law of Nature allowed fishing in the sea anywhere. Although she was setting something of an example by allowing foreigners to fish her own shores, principally because the country needed the fish, this attitude was nevertheless contrary to world opinion.

At this time Denmark claimed most of the North Sea and waters north of the Shetlands, Sweden all the Baltic, France most of the Mediterranean, Venice the Adriatic, Spain the Pacific and Portugal the Atlantic south of the Tropic of Cancer. Far from claiming exclusive rights to their own sea fisheries, English kings had entered into a series of treaties resulting in equal rights of foreign fishermen in British waters. In 1274, when twenty-two Flemish fishermen went up the River Tweed after salmon, they were seized and thrown into Norham Castle. Word got back to Flanders and Flemish fishermen killed scores of English fishermen (some reports say as many as 1,200 died). A few years later 1,000 men of France and Flanders joined forces to sack Great Yarmouth, but the town was forewarned and prepared a defence so the attack never came. The treaty Edward I negotiated with

7. *Wick Harbour at the height of its herring boom in the 1860s when 1,700 undecked boats took part in the six-week fishing season, crewed mainly by landsmen, and the port had the atmosphere of a frontier gold-rush town.*

Flanders forbade English subjects from molesting foreign fishermen and gave them the right to buy what they needed at Great Yarmouth.

The Scots were much more dependent on herrings as a basic food and were less tolerant of foreign fishermen, the English included. In 1532 several ships from Leith captured some Dutch busses, whereupon the property of Scottish merchants in the Low Countries was seized and for nine years Scots and Dutch were at loggerheads with neither side doing much fishing. On the union of England with Scotland in 1603 James I brought to the British throne the Scottish concept of exclusive coastal fishing rights. By joining the main headlands on a chart of the coast he enclosed twenty-six bays and estuaries such as the whole of the Bristol Channel, The Wash, and the Thames Estuary, which were proclaimed as 'King's chambers' exclusive to Britain. Then he told foreigners to obtain fishing licences from London or Edinburgh. Realising that this was the thin end of the wedge the Dutch sent men-of-war to protect their herring busses and entered into protracted negotiations. James was wary of pressing his claims too severely, mainly because he had made the mistake of letting the Elizabethan navy go into decline. His most famous gesture was to prohibit foreigners from fishing 'within a kenning of the land, as seamen do take a kenning' (later fixed at fourteen miles). The Dutch responded by doubling their escorts, but this was also due to harassment by French privateers based at Dunkirk.

By 1634 Britain's inability to keep order in her own waters was both ludicrous and pathetic. Dutch and French attacked each other in British ports while local people were powerless to keep the peace or protect their own property. French privateersmen chased a herring buss right into Yarmouth harbour, killed several Dutch sailors hiding on the pier, and robbed their ship. Town bailiffs fired two guns without success, and when the town marshall called upon the French in the king's name to desist and be gone he was answered 'with unseemly gestures and scorn'. Next day the privateer hovering off the coast was herself trapped by two Dutch warships and the crew ran her up the beach at Lowestoft so they could escape but as they waded ashore they were arrested and thrown into Yarmouth prison. In Scarborough the following year several citizens were hurt by flying bullets and houses damaged by cannon-fire when a Dutch man-o'-war chased a Dunkirker into the harbour. Twelve Frenchmen were killed and the rest swam ashore. Two weeks later a similar incident happened again in the same town; a month later at Blyth, Northumberland, a Dutch warship landed fifty men armed with muskets who marched in military order

8. *Newhaven (Edinburgh) fisherman and three boys, c.1843, portrayed in this very early calotype by David Octavius Hill and Robert Adams.*

through the town to attack a privateer in the harbour, and the men pursued the French for two miles into the countryside.

Charles I took sterner measures against the 'Hollander tigers', sending warships to collect payment of two shillings on every last of herrings caught. The king claimed to have collected 20,000 florins, but accounts which came to light much later suggest the amount was nearer to £501 15s. 2d. By illegally levying 'ship money' to build a naval fleet Charles found himself in deep constitutional trouble and during the civil wars that followed the Dutch enjoyed their last few years of prosperity from English herrings.

Cromwell's Navigation Act of 1651 and the power of his new navy spelled the first real trouble for the Dutch. The Act arose largely from fisheries disputes and stipulated that no fish (or other goods) could be brought into the country, or exported, or even carried from one port to another, except in British ships. For the Dutch cargo vessels, which had a virtual monopoly of the coastal trade, it was a death blow. But Holland had become so strong from the profits of fishing and trade that her resentment boiled over into war. During the first Dutch War, (1652–4), and the second, (1665–7), the entire Dutch fleet of herring busses had to stay at home, causing great economic hardship. The third Dutch war (1672–4) ended in a united front of both countries against France, but although Dutch vessels continued to land at Yarmouth every autumn for the herring fair, Holland's North Sea gold mine went into a decline. In 1703 the fleet with four escorting warships was attacked in the Shetlands by six French. When the flag-ship was sunk the remaining escorts fled and the French burned 400 herring boats. By 1736 the Dutch fleet was reduced to only three hundred sail and in 1779 to half that. The Dutch government tried to revive it by introducing bounties, but her once-great herring fishery did not recover for more than a century.

Although the Dutch virtually disappeared from the coast more than 150 years were to pass (after the Navigation Act) before British herrings secured a reputation as the best in the world. Throughout this period fishing was the despair of the Crown. Following the example of the state-controlled Dutch fisheries, successive British monarchs formed royal fishery companies subscribed with huge amounts of money from the exchequer and from investors, but each one was a failure. In Scotland, where herring fishing was a matter of subsistence rather than commerce, various kings had done everything in their power to encourage fishing but without result. In 1599 James I sent a colony of adventurers from Fife to Stornoway, but the islanders 'being rude and unpolished and the most dangerous and worst people' murdered many of them and the survivors fled.

The fishing companies launched by Charles I and II were granted financial support and special privileges such as the right to raise money by lottery and to collect in churches. On paper, the profits looked huge. Brand new and fully equipped for fishing a Dutch buss cost £500 and lasted for twenty years:

Costs	£
100 lasts of barrels	72
Salt	88
Beer	42
Bread	21
Bacon and butter	18
Peas and billets	6
Wages for four months	88
Wear and tear of ship and nets	100
	435

Income	
100 lasts of barrels filled or sold for £10 a last (Dutch prices were often much higher)	1000
Profit in one season	565

Shoals of herrings in the inlets of the Hebrides were so thick, according to the Rev John Buchanan in 1794, that 'the strongest whale dares not pierce through them, seeing he could not move his fins for the immense throng, and therefore browses behind the herring like a horse eating at the face of a hay-rick'. He remarked that if the Hebrides had been discovered in the South Pacific they would have been colonised by Englishmen at once. However every attempt 'to get the better of foreign fleets that have become a nuisance on the coast', and to capitalise on the gap left by their withdrawal, was a complete failure. Until the nineteenth century nobody investing in fishing made anything approaching a profit. The generally unsettled state of the country, distrust between English and Scots, the difficulty of finding experienced fishermen and the problem of marauding privateers accounted for the failures, but also there was overconfidence. Despite the example of the Dutch, herrings were still cured in imperfect barrels and ungutted, with the result that exported herrings were rejected. Houses were built for superintendents when the money should have been spent on cottages for fishermen, warehouses were constructed when fishermen had no boats. Queen Anne reorganised the various laws concerning herring in 1704 and offered to pay a bounty of £10 4s. on every last of herrings exported, and George I improved the offer in 1718, but reaction was negligible.

Typical of the way fishing failed is the story of the Society of the

Free British Fishery, incorporated in 1749 at a meeting held in the King's Arms Tavern in Exchange Alley, London. In varying amounts £214,900 was subscribed by 121 'gentlemen and merchants'. Experienced curers were engaged from Holland. Four busses were built, with a Dutch vessel as a model, and the fleet sailed in June, escorted by the war-sloop *Spy* to protect them from 'all insults'. At first the society prospered, capitalising on the new bounty of thirty shillings a ton paid on all fully-equipped decked boats that fished for herring between June and December, introduced that year by George II. In four years the society increased its fleet to thirty-eight busses, but the 230 sailmakers, shipwrights and coopers at its base in Southwold had difficulty getting official protection from press gangs, and nearly half the herrings sent to Hamburg were found to be bad and were not sold. Bounties were increased – to fifty shillings a ton, plus one shilling on every barrel sold for Scottish consumption and 3s. 4d. if brought to England – but in 1772 the society collapsed. Busses and other effects were auctioned at the Old Swan in Southwold, fetching £6,391 12s. 2½d. The investors lost ninety-two per cent of their capital.

Scotland's enterprise was a little more successful, but it was victimised by every kind of injudicious regulation and customs difficulty that Westminster could devise. When Charles II's last royal fishery company was dissolved in 1684 the magistrates of Glasgow took over its buildings at Greenock. Using small boats with crews of four, sufficient herrings were caught to meet local demand and provide several thousand barrels in a good year for export to France and the slave plantations in the West Indies. The small boats did not benefit from the tonnage bounties, and when the Scots built their own busses they were penalised by absurd regulations. English boats fitted out for the Shetland fishery could shoot their nets on mid-summer day, in accordance with Dutch practice, but those from the Clyde were prohibited from sailing until September 12, missing the best fishing and the best prices. Many new Scottish busses were built when the tonnage bounty was increased to fifty shillings in 1757, and while English vessels were paid on sight from general revenues the Scots were paid from customs, which was the only branch of revenue that fell short, so the fishermen seldom got their money. The Scots persisted, and despite fluctuations due to the brief withdrawal of bounties, the American wars which caused shortages of salt, pitch, and timber for making barrels, and the wars with France and Spain, at the end of the century there were 300 Scottish busses manned by 3,366 men and boys catching 60,000 barrels of herring a year.

Meanwhile George II approved 'An Act for the Encouragement of the White Herring Fishery' which permitted people to fish in any part of the British Isles, use any ports and harbours (unless artificially made) free of charge, and use all beaches and uncultivated land for one hundred yards above the high water mark to dry nets and land and pickle fish. The king also launched yet another fishing company, called the British Society, to organise a fishing industry in the north-west of Scotland. Instead of spending money on better vessels and nets, and providing houses for fishermen, the society repeated all the errors and misjudgements of the previous 150 years and built storehouses and residences for comptrollers and customs collectors at badly selected sites. No inducements or instructions were offered to local fishermen who adhered to their ancestral and uneconomic ways. Crippled by salt taxes and lack of transport between fishing ports and markets, crofter-fishermen led a miserable existence and in general fished only for their own benefit.

At the end of the eighteenth century the long years of mismanagement ended. The new century ushered in a period of spectacular growth and a sense of excitement that generated in Scotland the spirit of national endeavour that had been envied for so long in the Dutch. The first great herring fishery had founded the fortunes of the Hanseatic League, the second of Holland; now the third was budding – in Scotland.

While Napoleon locked Britain in war several events occurred almost simultaneously: the British Society launched a second scheme, improving the harbours and laying out complete suburbs for fishermen at Wick and Helmsdale; the Society of Arts offered a reward for the secret of curing herrings by the Dutch method; for a few years 360 large and 1,200 small boats profited from a superabundance of herring in and near the Firth of Forth, supplying valuable and wholesome food at a time when bread and meat were expensive because of the war. Herrings were sent directly to London by fast sailing smacks from Berwick; the scores of people salting and packing herring along the shore strongly resembled the Hansard fishery seven centuries before.

The 1808 Act 'for the further encouragement and better regulation of the British herring fishery' led businessmen to discuss Scottish herrings in the awed tones with which today they refer to North Sea oil. The bounty system was streamlined to three pounds a ton per vessel plus two shillings on every barrel of herrings cured and packed in accordance with regulations. A board of commissioners was appointed and fishery officers employed. Within a few years herring fishing was a major industry. As the Scottish boats were small the catch was brought ashore for curing, and this gave employment to hundreds of women as well as coopers, carters, and other tradesmen. The British Society's new scheme was a brilliant success and Wick became queen of the herring ports. The Society of Arts award was claimed by J. F. Denovan of Leith who was paid fifty guineas in 1819; he had hired six experienced Dutch curers to cure herrings on the west coast of Scotland and when the quality of his product attracted the attention of the authorities it set the standard adopted by the new Scottish Fisheries Board.

Herrings were no longer left on the beach for long periods but carefully shaded from the sun and gutted as soon as possible then packed carefully between layers of salt in barrels of specified size and quality. Where previously large fish got away because small-mesh nets were used, now only large fish were taken. The minimum weight of fish in each barrel was fixed by law and as a check on fraud and negligence a fishery officer branded each barrel with the date, the curer's name and address, and his own initials. The 'Crown brand' became famous throughout Europe as a mark of quality and merchants bought and sold thousands of barrels on the strength of it. Previously a season's catch of sixty to eighty crans of herrings per boat was considered fair fishing; now two hundred crans was less than satisfying and many fishermen caught twice that amount.

Bounties varied under the control of the Scottish Fisheries Board. At their highest, from 1815 to 1826, they were fifty shillings a ton plus four shillings a barrel. The sudden and spectacular growth of the Scottish herring fishery is evident from the figures. When the Board made its first bounty payments in 1809 fishermen cured 90,000 barrels of herrings. When bounties were phased out in 1830 the output of 440,000 barrels barely faltered. Fishing villages and harbours all round Scotland had been improved. Gutting and curing establishments were opened at Fraserburgh (1810), Helmsdale (1813), Macduff, Banff, Portsoy and Cullen (1815), Lossiemouth (1819), Burghead (1817), Peterhead (1820) and Lybster (1830); established ports such as Dunbar, Eyemouth, Burntisland, Buckhaven and Cellardyke also prospered. Larger boats enabled herring to be followed nearly all year, instead of only a short two-month season. Where the fleets travelled, so did the hundreds of wives and sweethearts who packed and gutted the herrings they netted. The season began in Shetland in May, worked down the east coast, and ended at East Anglia just before Christmas with important winter fishing available in the Firth of Forth; in Loch Fyne and the Clyde nets were cast whenever the shoals appeared.

In the middle of the last century a fifth of Scotland's herrings were landed at Wick: in 1820 the number of boats taking part in the midsummer herring season (mid July to the end of August) was more than six hundred; in 1862 there were over a thousand. During the six-week 'herring madness' the town was likened to Doncaster at the approach of the St Leger and something resembling a frontier gold-rush town. Men and women tramped across the Highlands to crew in boats and work in the curing yards; in 1840 the Rev Charles Thomson reported ten thousand young strangers in the town. Such an invasion brought its own problems. According to Thomson people slept ten or twelve to a room, typhoid was acute, the streets 'indescribably filthy – everywhere a putrescent effluvia steaming up from the fish offals.'

Herrings were supplied to curers by the cran, nominally one thousand fish but actually 1,320. The fish were measured out of the boat by the fishermen using quarter-cran baskets supplied by curers. These were universally larger than they should have been, so curers profited, but fishermen accepted the situation philosophically until the government introduced the Cran Measurement Act in 1908. In a wooden tank or bin known as the 'farlin' the fish were well sprinkled with rough salt which adhered to the scales and prevented the fish from sliding together when packed. Women worked in 'crews' of three, two gutting and sorting with lightning speed, and the third packing the fish into barrels, sprinkling each layer with salt. A barrel contained about seven hundred fish and a crew gutted and packed thirty barrels a day – a fish every five seconds throughout a ten-hour day. When full the barrel was fitted with a lid and left for ten days, then the pickle was run out through the bung-hole and the barrel was opened so it could be topped up. After the lid was refitted sufficient pickle was poured back through the bung-hole to fill any empty spaces.

The system of curing herrings on land, adapted from the Dutch who cured at sea, became known as the 'Scotch cure'. With this Scotland finally realised her potential as a fishing nation. By the end of the century catches and exports quadrupled and Britain's North Sea gold mine was the largest fishery the known world had ever seen.

The crown branded on the bilge (widest part) of every barrel of cured herrings which had been inspected amounted, in effect, to a government guarantee of quality. In addition, an official stencil was used to describe the barrel's contents (see picture). This gave the month and year of

Brass stencil used for 'branding' barrels of herrings cured and packed to specifications laid down by the government. The letters D.S. are the initials of the Fisheries Officer. This Crown brand became recognised all over Europe as an official seal of quality.

cure, the initials of the fisheries inspection officer, and the grade of herrings. These were:

La. Full Large herrings full of milt or roe, longer than $11\frac{1}{4}$ inches·

Full Large herrings full of milt or roe, longer than $10\frac{1}{4}$ inches.

Filling Maturing herrings with gut removed, longer than $10\frac{1}{4}$ inches

Mat. Full Full of milt or roe, longer than $9\frac{1}{4}$ inches.

Medium Maturing herring, gut removed, longer than $9\frac{1}{2}$ inches.

*Mattie** Young maturing herring, gut removed, longer than 9 inches.

La. Spent Spent fish (milt or roe absent because breeding has been completed) more than 10 inches long.

  * *From the Dutch word for maiden.*

Peggoty's Port.

'The fishiest town in all England' was how young David Copperfield described Great Yarmouth when he visited the place with Peggoty in Charles Dickens' story set in the early years of the nineteenth century. He would have liked it better 'if the land had been a little more separated from the sea, and the town and the tide had not been quite so mixed up, like toast and water.' But when he got into the street 'and smelt the fish, and pitch, and oakum, and tar, and saw the sailors walking about, and the carts jingling up and down over the stones' he felt he had done the place an injustice. Then he turned 'down lanes bestrewn with bits of chips, and little hillocks of sand, and went past gas-works, rope-walks, boat-builders' yards, ship-wrights' yards, ship-breakers' yards, caulkers' yards, riggers' lofts, smiths' forges, and a great litter of such places . . .'

For two centuries, from the Navigation Act of 1651 until the spread of railways, Great Yarmouth was one of the most important ports on the east coast because it offered the only secure anchorage between the Thames and the Humber. Five miles from its beach lay a long sand-bank, now much reduced in size, which offered reasonable shelter for masses of colliers and other cargo vessels navigating the low-lying and hazardous coastline. It was common to see a thousand ships anchored in Yarmouth roads awaiting favourable winds and in the 1860s, when railways began to take over from shipping, it was reckoned that 50,000 ships (not counting fishing boats) passed Yarmouth every year. Travelling north from Yarmouth by land about 1718, Daniel Defoe noted that farmers 'had scarce a barn, or a shed, or a stable; nay, not the pales of their yards, and gardens, not a hog-sty, not a necessary-house, but what was built of old planks, beams, wales and timbers &c, the wrecks of ships, and ruins of mariners' and merchants' fortunes.' Defoe reported a tragedy that occurred in 1692 when a fleet of empty colliers heading north, and another carrying coal southwards, were both driven into The Wash and unable to beat clear, 'so that in the whole above two hundred sail of ships, and above a thousand people, perished in the disaster of that one miserable night, very few escaping.'

After six centuries the autumn herring fair ended finally about 1800, but its demise had been signalled as early as 1663 when Yarmouth Corporation at last got rid of the jurisdiction of the wardens of the Cinque Ports and prohibited foreign fishermen from selling herring. To monopolise the trade all boats except those of Yarmouth had to sell to freemen of the borough, a ring of about fifteen merchants who named their own prices. In 1709 these restrictions were ended but the damage was done and the herring fair never fully recovered. During the eighteenth century small Dutch fleets of about fifty herring boats, then called schuyts, sailed up the river in line. On the Sunday before Michaelmas Day they were decorated with flags, and people came from as far away as Norwich to take part in the small quay-side fun-fair and listen to the Dutch fishermen, in their wide breeches, singing songs in the moonlight.

Throughout the Dutch dominance of the herring fishing, and the limbo that preceded the Scotch cure, Great Yarmouth's main resource was the smoking of red herrings which were sent all over the country and exported in open-stave barrels to Mediterranean countries, particularly Spain. Legend has it that a poor fisherman found himself with a surplus of herrings which he hung in the rafters of his smoky hovel and forgot about them. When he brought them down a fortnight later the fish had been smoked and cooked until they had the appearance of polished mahogany and a strong smoky flavour that went admirably with such things as chewing tobacco and strong ale. In the sixteenth century Thomas Nashe described the 'high-dried' or 'ham-cured' herring as 'wholesome on a frosty morning: it is most precious merchandise because it can be carried through all Europe. Nowhere are they so well cured as at Yarmouth. The poorer sort make it three parts of their sustenance. It is every man's money, from the king to the peasant.'

The advantage of 'redding' herring was that the cheapest fish could be used because they did not have to be in prime condition. This enabled boats to stay at sea for several days, fishing until their holds were full. Also, the cure required little expensive salt and it had to be done ashore, which gave Yarmouth curers an advantage over the Dutch who were far from home. At the end of the eighteenth century Yarmouth fishermen had a fleet of about fifty handy and very swift three-masted luggers. Another thirty or forty fished from Lowestoft,

9. The Rows of Yarmouth, seven miles of criss-crossing narrow alleys, were fishermen's homes until early this century; sea breezes kept the streets pure and despite the cramped quarters living conditions were not unhealthy.

1639 A YARMOUTH ROW.

then a small haven that had long been a jealous and less prosperous neighbour ten miles south of Yarmouth. Later the luggers carried only two masts which were 'duffed' or lowered when drifting to their nets. The herring season lasted from the middle of September until the middle of December. With crews of about twelve hands, the luggers stayed at sea for three or four days, rough-salting the catch on board and landing it for smoking, making returns to the Customs House to claim their bounties. Lowestoft red herrings commanded slightly higher prices in the home market because the smokehouses were situated in narrow streets and passages called scores under the town's cliffs which were said to affect the draught and produce a better class of smoked fish.

At other times of year the quaint luggers landed mackerel and line-caught fish on Great Yarmouth beach, and great was the excitement as beachmen pushed off in their 'bullock boats' or ferries to meet them. After the scramble to get the catch ashore the fish were gutted by fishwives sitting on boxes and packed in wicker peds, or hampers, the largest fish arranged on top. A road surface of tanners' bark (used to treat nets and sails) was laid across the soft sands for the troll-carts which collected the fish after it had been auctioned and trundled it away to be loaded into horse vans that rushed it to London overnight.

The troll-carts were very narrow, two-wheeled, low-slung vehicles, similar to small chariots, which were built for the 'rows' where most fishermen and mariners lived. These were a curious gridiron of 145 cobbled lanes, in some places so narrow that on meeting another pedestrian you had to press into a doorway to pass. Kitty Witches Row, fifty-six inches wide at one end, twenty-nine inches at the other, and only twenty-seven inches in the middle, claimed to be the narrowest street in the world. Upper floors of houses overhung the lanes and neighbours could shake hands across the street. Others joined over-head, forming a tunnel. Here there was only a blank wall, there a tiny courtyard where washing and starching were done, and nasturtiums and scarlet runners grew in boxes.

The origin of the rows, and their name, is obscure. The criss-cross network of streets, sloping towards the river and open to clean sea breezes, might have been based on a fishing-net pattern, or perhaps the streets were laid out on ground used to dry nets, where there were narrow footpaths between the different plots on which nets were suspended from poles. The most likely explanation, however, is simply pressure of space, for until the middle of the eighteenth century the town was enclosed by its original walls. The name is a likely corruption of the French *rue*, meaning street.

The seven miles of rows contained 1,811 houses. Although each row had a number it also had one or more local names such as Money Office Row, Adam the Barber's Row, and Budds' Sailmakers' Row.

The little streets were never still; even in the dead of night there was movement, as skippers went out to their vessels and the watchman patrolled, crying the wind – 'East is the wind, east-north-east; past two on a cloudy morning!' There was poverty in the rows but little squalor. The houses were nicely kept, with flowerpots in the windows. Whitewash was laid on liberally in every corner and the children were healthy and well turned out. There was always a smell of fish, however, because most people earned their living in fishing and some boiled shrimps in their houses, but it was said that of the seventy-two foul odours the poet Samuel Taylor Coleridge listed in Cologne, not one could be found in the rows of Yarmouth. Here you could find almost anything, as the magazine Household Words reported in 1860: 'If you want a stout pair of hob-nailed shoes, or a scientifically oiled dread-nought, or a dozen bloaters, or a quadrant or compass, or a bunch of turnips, the best in the world, or a woollen comforter and night-cap for one end of your person, and worsted overall stockings for the other, or a plate of cold broiled leg of pork stuffed with parsley, or a ready-made waistcoat with blazing pattern and bright glass buttons – with any of these you can soon be accommodated in one or other of the rows.'

Cod Bangers.

Demand for fresh fish in the fashionable quarters of eighteenth-century London evolved a new technique for supplying fish that originated, like most other significant developments in the early days of the fishing industry, in Holland. Whole cod's head served with oyster sauce was a popular West End dish but prime fish with the fresh tang of the ocean deeps in its flesh was rare and prohibitively expensive. To take advantage of the high prices fishermen at Harwich and Barking in about 1712 adapted a Dutch idea: as soon as they were taken off the hook fish were put into a 'well' filled with sea water aboard the fishing smack. The well comprised two water-tight bulkheads across the whole width of the hull. The planks between them were drilled with holes so that sea water washed in and out. Fish in the well did not have to be fed and remained alive for several weeks. The fishing boat could stay at sea until its capacity was filled, then head for the market with no risk of the catch deteriorating due to delays caused by head winds. If there was a glut at the market the vessel could stand off and wait for the price to improve. By the middle of the eighteenth century Harwich had a fleet of sixty-two well smacks fishing in the North Sea and supplying London with fresh fish landed directly at Billingsgate, but its importance declined in the 1790s with the rise of Barking.

Barking was then only a remote village two miles up a tidal creek but as a fish supplier it was the premier fishing port in the country, the forerunner of Hull and Grimsby. At the end of the Napoleonic Wars it was the home port of about 130 first class fishing smacks built stoutly of Essex oak which had been matured for at least twenty years. Its vessels were chartered by the navy for use by press gangs collecting seamen, and as tenders for men-o'-war. A 1766 newspaper advertisement described Barking's handy and seaworthy vessels as 'the fastest shipping that swims.' At its zenith in 1851 the little port was the home of the biggest fishing fleet in the world, 220 first class fishing vessels sailing under the plain blue square house-flag of Samuel Hewett and known from the flag as the Short Blue Fleet. The fleet had been started in the 1760s by an orphan from Fife, Scrymgeour Hewett, who married a smack-owner's daughter and settled there. Samuel Hewett, his second son, was a man of adventure who ran away to sea, was apprenticed as a smack's boy, and became a great pioneer of North Sea trawling.

A typical well smack in which Samuel Hewett might have sailed as a boy was about sixty feet long, cutter rigged, with an easy motion. The well was contained within two massive caulked bulkheads, about fifteen inches thick and twelve feet apart. These had to be completely water-tight and stronger than any other part of the vessel because the space between them was filled with water. Holes to let the water in and out were augered low down near the keel, so that when the vessel came upright after heeling the air trapped inside the well did not force the water upwards through the grating-covered hatchway, or 'funnel'. As flatfish covered the holes if they were allowed to lie on the bottom they were kept in perforated boxes, or trunks, which were lowered into the water. Big fish such as halibut and skate were tied by the tail or by the nose and suspended from a wooden rod beneath the water.

Every day fish in the wells had to be inspected to ensure that none had died, or had been damaged by bumping against the bulkheads as the smack pitched and rolled. To climb down into the dark and slimy pit of the well, with water surging to and fro and five hundred large free-swimming cod in a solid mass like giant maggots in a tin, must have been a horrific experience.

Until 1770 fish were caught with hand-lines: two hooks on a stout line carried to the bottom with a seven-pound lead. Then another idea was copied from Dutch fishermen seen working the North Sea. This was the use of great-lines ('grut-lines') or long-lines, which were laid along the sea-bed by the mile and marked with buoys then collected later. A typical cod smack carried sixteen dozen lines, each thirty fathoms long. About every seven feet there was a short piece of line attached, about thirty inches long, called a 'snood'; at the end of each snood was the baited hook. A train of lines stretching six miles, held by an anchor and marked by a buoy every mile, carried five thousand hooks; by the 1860s lines were ten or eleven miles long, but were not always set in one length. Large quantities of bait were required, at considerable expense. A cod smack sailed with forty wash (210 gallons) of whelks in net-bags in its well and in the 1860s at Grimsby alone forty small cutters were engaged exclusively in collecting bait from The Wash.

Lines were shot at sunrise, always across the tide so the snoods streamed away from the line. It was a tricky business, deftly paying out the coiled line and the thousands of baited hooks as the smack sailed along at four or five knots. The lines had to be hauled in before sunset and most cod smacks carried a roomy small boat, eighteen feet long and with its own well, for doing this. Big cod and other 'prime' fish

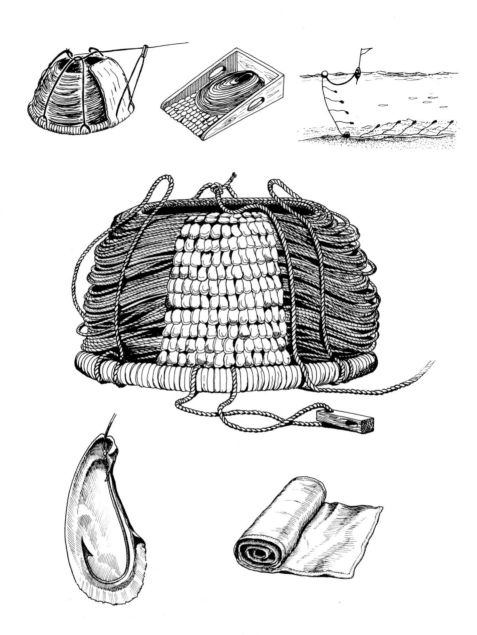

Fishermen's wives, mothers, sisters and daughters collected shellfish with which to bait the long-lines used in cod fishing.

hooked in the lip, with no damage to the throat, were dropped into the well (tail first so they didn't break their necks). The remainder were gutted, split and laid down in salt or ice.

North Sea cod is better eating in winter, so it was said to be not fit for the table 'until it had snow in its mouth'. The smacks fished local waters at the southern end of the North Sea from October until about March, then fitted out for three or four months of fishing in the Faroes and Iceland. During the Iceland season a well smack made about three voyages, landing a total of twenty thousand good cod. During the first two voyages the fish were split and salted and landed at the Shetlands or north Scotland, where fast carriers collected the fish for Liverpool which had a flourishing export trade to Mediterranean countries. On the third voyage the fish were kept alive in the well, first having their air bladders pierced with a 'cod pricker' made of a sharpened silver spoon; the silver prevented infection, the pricking enabled the fish to swim near the surface. Sailing homeward the codmen hailed inshore fishermen on the east coast for any turbot they might have, and money to buy it was carried by the skipper in what was called the turbot bag. The crews of between seven and fifteen hands were usually on wages, but many were apprentices who were paid fourpence a week pocket money and twopence a score for every fish caught by hand-line which measured at least four planks across the smack's deck; when landing a cod the boy cut the barbel or 'feeler' from under the jaw to keep tally of the number he had caught.

The cod smacks sailed to Gravesend where the live fish were taken from the wells by long-handled nets and put in seven foot long floating chests moored in the river. These were open between the planks, with padlocked lids; the ends were pointed to present less resistance to the tide. According to demand at Billingsgate the cod were ferried up the Thames by small sailing 'hatch' boats where they were killed and sold, ultimately to reach the fishmonger's slab in the West End as 'live cod' and fetch as much as a pound each. In the 1830s, when the Thames became too polluted, the hundreds of cod chests at Gravesend were transferred to Harwich which supplied Billingsgate direct by way of the new railway.

Well smacks survived the sudden rise of trawling in the middle of the last century because they provided small quantities of the highest quality fish which earned the best prices. Although long-lines were occasionally dragged away by trawls, there was basically no conflict because the lines were usually set on rough ground where trawls could not go. The market for 'live cod' continued long after the use of ice and rapid transport by railway had made fresh fish available to ordinary folk in huge quantities. Well smacks did not begin to decline until about 1880, when there were three or four hundred chests floating in Grimsby dock, containing up to twenty thousand live fish.

When required a chest was taken alongside a barge and lifted up by a derrick. Then a man armed with a cudgel jumped into the chest among the cod as the water drained through the gaps between its planks, and whacked each fish on the head. Grasping it by the muzzle, he then tossed it at the feet of two other men who clubbed it again on the snout, so that the fish died limp rather than stiff. Further to lessen the toughening effects of rigor mortis the fish were 'crimped' – deep transverse slashes were made across the body to sever the skin and longitudinal muscles. Large skate were crimped while still alive, then plunged into fresh water. The 'live' fish were immediately loaded into special railway trucks which were attached to an express train and rushed to Billingsgate. Sometimes cod were kept in the harbour tied by their tails to a thick rope so they radiated outwards like a huge cartwheel, and when picked up looked like a wriggly and slimy bunch of green and silver onions.

Well smacks and the spectacle of dock-workers bludgeoning cod with pick-axe handles were seen less and less when ice was used for preserving fish at sea. The last well smack sailed from Barking in 1880 and from Grimsby about 1904. The skipper of *King Arthur*, one of the last 'cod-men', also happened to be one of the very few fishing skippers to keep a close record of his catches. Two extracts from the log of Edwin Green Smith show the hazards and rewards of long-lining in the North Sea:

On December 21, 1886, he sailed from Grimsby for Flamborough Head, wind from the south-west, weather fine. Next day he shot his lines in seventeen fathoms, with a fresh breeze blowing. He had hauled only three shanks of the line when it broke and he had to sail to find the other end of it, catching four cod. Before all the lines were up it blew very hard from the south and he had to abandon the rest, take two reefs in the mainsail and hoist the storm jib. Late next afternoon, after lying hove-to all day, he started to look for his lines, found them on Christmas Eve, and brought in a score of cod worth four pounds.

On June 14, 1887, 'as Soon as the Ship got Out of the dock E. G. Smith went up on the Mast Hed found Birds Nest in the Cap with 2 Eggs in'. Next day lines were shot in twenty-three fathoms and they

---

10. *Fish being ferried ashore to Great Yarmouth beach and loaded into troll carts early last century; prime fish was then relayed to London by fast horse van. Beach look-out towers flanking Nelson Monument served longshoremen who serviced the scores of sailing vessels that passed Great Yarmouth daily.*

11. *Well-smack recovering its long-lines in a rising sea; live cod in good condition were dropped into a midships part of the hold where the hull was perforated to let the sea flow through.*

got 120 cod, a ling and a skate, in fine southerly weather. Another 60 cod were caught on the following day. On June 17, the weather still warm and southerly, they took 90 cod and on reaching the dock next day their catch fetched £17.

Silver Pits .

With a supply of quality fish assured for the high-priced markets like the West End, and the tremendous growth of population and commerce that followed the peace with France in 1815, what the fishing industry now had to supply was quantity. It was the devilish wondyrchoun, that had caused such a lot of trouble three and a half centuries before, which filled the need. The power of sail was now sufficient to haul a large trawl along the sea-bed, scooping up fish as it went. It was developed simultaneously in Barking and Brixham, and both claim credit for its invention. It seems likely that Barking men adapted the idea very early from the oyster dredge used by the Romans. But Brixham men took the idea from Dutchmen, who came over with

Beam trawl. The length of the beam and size of the trawl depended on the length of the smack. The biggest beams were fifty-two feet long, but forty-eight feet was more common.

William of Orange when he landed there in 1668. The Brixham trawl was larger than that used in Essex, the mouth of the net being held open by a thirty-foot beam of elm which slid over the ground on curved iron runners like those of a sled. With the bright, clear water and the enormous swells of the open Atlantic on their doorstep, Brixham fishermen developed vessels and fishing gear of great strength while Essex men, fishing a shallow sea of cloudy waters, hidden shoals and short, steep waves achieved much the same result for different reasons.

Brixham in Devon had been a fishing port at least since the Domesday Book, supplying fish inland to Exeter, Bristol and Bath. It is probably unique in that for more than a thousand years it has always been a fishing port, never a commercial one, and even in recent times has survived the tourist boom with its identity and fish quays modified and modernised, but intact. In 1785 the picturesque little harbour already had seventy-six large decked trawlers, each of twenty to twenty-five tons. Every day fish were sent inland by horse van but the cost of transport was great and it was only the shortage of fish during the Wars with France that allowed it to be competitive. With peace in 1815 the price dropped alarmingly so Brixham trawlers moved eastwards to be nearer London, sailing from ports such as Dover and Ramsgate. Fishing banks in the Straits yielded rich hauls of sole that could be sent by fast mail to London where it became a popular and fashionable dish.

Soon Devon trawlers were competing with Barking trawlers in the North Sea, and every summer a small fleet began to base itself at Scarborough where the fashion for sea-bathing provided a market of hearty appetites for sole, turbot and other flat fish trawled from the near-shore banks. Many Brixham skippers took their wives and families for the season, settling them into temporary homes at Ramsgate or Scarborough.

In the early 1830s trawling was poised on the brink of creating a North Sea boom that would be greater and more important in the long term than the herring fishery born only a few years earlier in Scotland. At this time rapid transport by railway had extended to only a few points along the coast. Fish could be preserved at sea only in small numbers, in well smacks, and fresh fish was still the privilege of the wealthy. The Newfoundland cod fishery was all but finished, having declined from 300 English ships in 1792 to only fifteen in 1823, mainly because of competition from Americans and Canadians.

There are several versions of the discovery of the Silver Pits. The most likely one is that in 1837 three or four Brixham trawlers supplying Scarborough were dispersed by an early winter storm. All but one, of which the skipper was probably William Sudds, a Brixham man who had settled in Ramsgate, reached Scarborough safely. Sudds limped in two days later. One account says his beam was 'clean', the net having

74

been torn away by weight of fish. Another described the smack staggering in under the weight of more than two thousand pairs of soles caught in a single haul. Sudds had marked his spot and returned for yet another fabulous haul of soles, but the secret could not be kept for long and soon it had swept the coast. From Brixham a fleet of smacks immediately set out for the North Sea, those who knew the route escorting those who did not.

What Sudds had found – and was subsequently named the Silver Pits – was a long and narrow gut, or under-sea valley, twenty-four miles long and between one and two and a half miles wide. It ran from north to south, its north end twenty-five miles from the mouth of the Humber. To this deep area, and many others like it, soles were driven to avoid the cold water issuing from rivers such as the Rhine. Soon afterwards the concentrations of fish dispersed, to occur again during severe winters which became known to trawlermen as 'pit seasons'. The spectacularly rich fishing that followed Sudds' discovery focused attention on the potential of the North Sea, and in particular on the vast submarine plateau, 6,800 square miles in extent, known as the Dogger Bank.

Well smacks and line fishermen had fished the Dogger Bank for decades. Now, dragging trawls which enabled them to explore the hidden landscape like blind men with sticks, trawlermen learned every twist and hollow of the gulleys and pits in and around the great bank, feeling their way by touch and sometimes gripping a trawl warp with their teeth, putting their fingers in their ears, to sense the nature and lie of the sea-bed by vibrations travelling up the rope. By casting a lead-weight armed with tallow, which told them the exact depth and brought up samples of the bottom, they could identify their positions exactly. A cast that brought up sand from thirty-two fathoms put them over a hollow called Brucey's Garden; Markham's Hole showed mud at thirty to thirty-five fathoms; Botany Gut, an inlet in the South Rough, near the Oysters, showed mud at forty fathoms. There was no formality in the naming of these undersea areas: the names just 'grew', like The Cemetery and The Hospital, two notorious fishing grounds on the east face of the Dogger where the bank formed steep underwater cliffs that caused very dangerous seas. And all the time the fishing was marvellous. Here was what the markets required – quantity. Previously, at Brixham, two tons of fish was a good haul. Now, anything under five or six tons was not worth mentioning.

But finding and catching the fish was only half the problem solved. The other half was getting the fish to the markets that required it. Scarborough was a good market during holiday time but for the rapidly increasing numbers of smacks fishing the North Sea it was unsuitable as a home port. The small harbour dried out at low tide and the anchorage offered no shelter from north-easterlies. Ashore there were no facilities or stores, and the local line fishermen were hostile to the invading trawlers. The adventurous West Countrymen needed to be nearer to the new fishing grounds than their forward camp in Ramsgate so many of them loaded furniture and belongings aboard and migrated to Hull.

A busy commercial port that rivalled Liverpool, and sufficiently go-ahead to employ one of the first steam tugs, Hull did not welcome the fishermen. This attitude was curious because the city had no fish market and the shortage of fish for its own residents had been a problem for years. Shrimps were trawled in the river and oysters grew in great quantities along the shore, but apart from sprats which were sold as fertiliser by the ton the fishery in the Humber was meagre. The city even offered its own bounties – ten pounds to the fishermen who brought the largest quantity of fish by sea or overland during 1793 – and later spent as much as £1,500 a year on inducements to fishermen to bring fish to Hull.

Yet, when fishermen wanted to make use of Hull's docks with its railway and facilities such as shipwrights and chandlers, the authorities could not have been more difficult. In 1845 there were twenty-nine smacks belonging to the port, including William Sudds who settled there, and with seasonal visitors the total fishing fleet numbered about forty vessels. In ten years the number had increased to more than 100, but they were still regarded as an intolerable nuisance and there were no facilities of any kind. The railway station was more than a mile and a half away from the quay where there was room for four smacks to berth. There was a single shed on which somebody facetiously daubed the name Billingsgate in big white letters. Other boats had to put their fish in baskets and get alongside the quay as best they could in small boats between one steamer's bow and another's stern. There was no quay space for stores, no berths for refitting, no dry dock. To scrub their vessels' bottoms fishermen had to moor alongside the promenade and work when the tide fell, but the position was exposed and vessels bumped heavily against the wall in the surge and were often damaged. The fishermen, however, persevered. The Humber was the most important waterway north of the Thames and offered a safe approach in bad weather with a sheltered winter anchorage in the lee of Spurn Point and this was an incalculable advantage over places like Scarborough, even if a sail of at least twenty miles requiring intricate pilotage in a dangerous tideway was necessary to reach the docks.

Discontent at Hull was an unexpected bonus for the railway companies which purchased an old dock in a creek on the south shore of the river and situated seventeen miles nearer to the sea – Grimsby. In the Saxon period it was an important trading port but the haven silted up and the town suffered a serious depression until a company of merchants built a new dock in 1800. A large part of the Baltic timber trade was won from Hull and by the end of the French wars its population had trebled, although with 60,000 people Hull was still twenty times larger. Its genesis as the metropolis of the fishing industry came in 1848 when the Great Northern Railway, having acquired the docks, opened lines that linked the port to London and the burgeoning industrial cities of the North and Midlands. In 1852 the Royal Dock was opened with comprehensive facilities for fishermen, including a complete dock for their sole use. Nine smacks from Barking, three from Hull and two from Scarborough joined four trawlers acquired by the railway company. West Country fishermen searching for a base but dissatisfied with Hull were attracted by low dock charges and the town corporation greeted the influx of settlers by laying out new streets and leasing house-building land cheaply. In 1854 the railway despatched 453 tons of fish and fish merchants were offered free tickets for the purpose of drumming up new business. By 1860 the tonnage had increased ten-fold and there were 315 smacks at the port. Seven years later a fish-landing pontoon 3,000 feet long, two fish docks twenty-three acres in extent, and a graving dock that held ten smacks at once had been built. Allied trades became established – there were now scores of sailmakers, shipwrights, smiths, coopers, blockmakers, twine spinners, pontoon labourers, and smoke-houses. By 1877 Grimsby was a town of 33,000 people of whom two-thirds were involved directly or indirectly in the fish trade.

Hull responded to Grimsby's competition and matched its meteoric development of fisheries, although in 1877 only one-fifth of its population depended on the fish trade. When fishermen were given Albert Dock to call their own, in 1869, there were 576 smacks registered in Hull and 242 others used the port regularly. Sail-barking yards, ice-houses and smoke-houses were built around the dock but soon it was so crowded that smacks often had to anchor in the river for four tides before they could enter. In 1883 St Andrews Dock was built three miles further up the river, to the dismay of fishermen who now had to work a perilous way through dozens of merchant ships and barges anchored in the roads, and also had to remove or abandon their dock-side facilities. However the new dock was soon established as a permanent home of the fishing fleet and it remained so until a new dock was built in 1976.

Trawlers also colonised Great Yarmouth and Lowestoft. The fleet grew from about a dozen in 1850 to more than sixty in five years when numbers were further swelled by the entire Short Blue Fleet which Samuel Hewett removed from Barking and settled at Gorleston, on the south bank of the Yare opposite Yarmouth. By 1875 about 400 trawling smacks sailed from the port, in addition to a large fleet of herring drifters supplemented in autumn by several hundred Scottish drifters. In Lowestoft, Great Yarmouth faced the same kind of com-

77

petition as Hull did with Grimsby, for this was another small haven rescued from depression and developed for the fishing industry with sidings and new docks by a railway company. Lowestoft became a herring port, and a base for trawlers known as 'single-boaters' which worked alone rather than in fleets. Very little trawled fish was landed at Great Yarmouth, but it was the home base of at least four big trawling fleets.

The fleet system was introduced early in the 1850s, probably by Samuel Hewett, for the sake of efficiency. When it was a particular smack's turn to leave the fishing ground for a stay in port, after about six weeks' work, its skipper hoisted a flag and the night's catch of the whole fleet of a hundred sail was brought aboard by small boats. Each vessel also sent a fish-note giving particulars of its catch. On arrival in port the smack moored beneath a steam crane and the fish were hoisted out, the skipper handing over the fish notes, or 'pot list', so the salesmen knew to whom the fish were consigned. Fleeting spelled profit, because smacks were able to quarter the whole of the North Sea without the necessity of sailing back and forth. But it had a profound effect on the lives of smacksmen because it meant long periods at sea and short stays in port; a whole generation of fishermen spent a lifetime almost continuously at sea, with only six weeks in port every year. Hull smacks fished in a single fleet from April to September, but during the summer months fish tended to deteriorate during the sail homeward so they joined the Short Blue Fleet in which Samuel Hewett had also pioneered the use of ice.

The idea of using ice had come from China in 1786 when an English traveller called Alexander Dalrymple reported that the coasts abounded with 'snow houses' and Chinese fishermen carried snow in their boats which enabled them to carry their fish far inland. A Scottish country gentleman called George Dempster applied the idea to salmon, which until then was eaten fresh only locally, and was mainly 'kippered' or smoked to preserve it. Early in the nineteenth century ice houses were built at the chief salmon centres, such as Berwick. Fresh salmon was packed in ice and carried by fast sailing packet (and later by steamer) to the London market where the demand at once led to a huge rise in prices and salmon became a costly commodity.

Samuel Hewett realised Barking was surrounded by marshes which froze over in heavy frosts. Watchers were employed to patrol the ponds to ensure that the ice was not disturbed as it formed, then it was cut and delivered to a special ice cellar that had walls eight feet

12. Topsails struck preparatory to entering Lowestoft and a good reef in her main, this trawling smack (note the beam trawl lashed to the port quarter) flies along at an easy ten or twelve knots.

thick and a capacity of ten thousand tons. In storage the thin sheets of ice stuck together under pressure and formed blocks which remained frozen throughout the following summer until November. Farmers received ten to fifteen shillings for a 30cwt cart-load early in the winter, declining to about four shillings nearer spring, and for many it was the year's most valuable crop. Small boats called ice lugs carried ice down the creek to supply trawlers and cod smacks, which now could dispense with their wells. For the Hull smacks making use of his ice, Hewett made a service charge of twenty per cent. In 1859 Hull smack-owner Robert Hellyer, who had migrated from Brixham four years earlier (and whose family firm is now the largest trawler company in Britain) had the idea of importing ice from Norway.

Four Hull owners clubbed together and built cutters, larger than ordinary trawlers, for collecting the ice. Cellars of a large warehouse near the dock were rented for storing ice in sawdust. The ice was cut from frozen lakes in Norway using horse-drawn ploughs and wide-toothed saws, and was more hygienic than English pond ice. The

Blocks of ice cut from frozen lakes in Norway were packed in sawdust and brought to Grimsby and Hull by barque.

scheme was immediately successful. In Grimsby the dock company built an ice house with a thatched roof which was supplied from surrounding ponds and topped up with ice from Norway. By the end of the 1860s thousands of tons of natural ice was brought to the fishing ports every year in large wooden-hulled barques. One of the earliest consignments to Hull suffered a disaster when a storekeeper left a valve open and the cellar was flooded by the tide, ruining a whole year's supply.

The fleeting system was given impetus at this time by onerous railway freight rates; also the companies refused to return empty fish boxes at a nominal charge. Hellyers and other trawler companies in the Humber formed their own fleets and built big, fast sailing cutters to service the fleets and take their catches direct to Billingsgate. These superb vessels were the fastest on the coast, and when they took time off from summer fishing to collect fruit from the Mediterranean they showed their heels to speedy Salcombe fruit schooners in anything but a following wind. But their speed depended entirely on the wind, and when calm weather found them ghosting up the Thames against the tide with holds of fish in danger of turning rotten, steam tugs were sent out to look for them.

In 1865 fish carriers powered by steam were built, first for the Short Blue then for other fleets. These vessels, which continued to be known as cutters, were the hustlers of the North Sea, an ever-running link between the London market and the fleets of smacks toiling far out at sea. They were long and lean fliers, built to drive through any sea with four thousand trunks of fish packed in crushed ice in their holds, decks awash and the helmsman crouching at his wheel behind a canvas dodger. The steam cutters carried trawls so they could be classed as fishing boats and pay smaller port dues. The sailing cutters were converted for trawling, their immense rigs being divided to make them more easily handled as ketches. They proved so successful that many conventional fishing smacks were lengthened by ten feet and given the new 'dandy' rig.

Little more than twenty years after the discovery of the Silver Pits in 1837, North Sea trawling was an efficient and highly organised industry employing nearly two thousand fishing smacks and supplying all of industrial England with daily consignments of fresh fish.

The Great Emporium.

The effect of railway transport in the middle of the last century was to concentrate the fishing industry in a smaller number of ports fortunate enough to be linked to the network. Ports with no railway found themselves out on a limb, their manpower being attracted towards the more flourishing harbours. Those with difficult connections between harbour and railway, such as Scarborough and Brixham where fish had to be conveyed by cart through steep streets, were at a grave disadvantage compared with railway-owned ports where sidings were adjacent to the fish docks.

The daily carriage of fish throughout the country was so important that principal railway companies had special departments to run it. In Grimsby, trains were loaded four-abreast; each ventilated fish van took fifteen tons and had a concrete floor sloping towards a drain-hole in the centre. Between the first departure at 9.09 a.m. (for West Riding and Scotland) and the last at 8.20 p.m. (for Lancashire and Yorkshire) Grimsby despatched thirteen fish trains which had precedence over some passenger and all other goods traffic.

It was transport, but only at a price. In the 1880s the cost of sending fish by rail was an average forty per cent of its gross wholesale value, and sometimes as much as sixty-eight per cent. A fisherman had to think twice before consigning fish by rail because he could be out of pocket if it did not fetch a good price on arrival. It was known for Scottish fishermen to send herrings by train to London and receive as their profit on the venture only a penny stamp. And all fish was sent at the sender's risk.

The railway companies were outrageously unfair in their loading of freight rates for fish. A cran of herrings cost one pound to send from Liverpool to London, but only two shillings from Liverpool to Germany. A Frenchman could send a barrel of herrings via Dover to London for 1s. 8d., but it cost a Folkestone man 5s. 6d. Norwegian fish was brought to London for fifteen shillings a ton, but Scottish fish cost £3 15s. Fish delivered to Liverpool from Scotland by steamer cost

32s. 6d. a ton to rail to London, but fish imported from America and other countries cost only 25s. The result was that some of the big fleets such as Morgans broke up during the 1870s because the owners could not make a profit.

Rapid transport had a curious effect on local markets. Mackerel brought by train from Cornwall for sale on Brighton beach could fetch higher prices than mackerel landed that day by Brighton fishing boats. The reason was that local mackerel was larger and therefore less favoured by the town's hoteliers, and also had probably been on ice for several days, while Cornish mackerel had still been swimming less than thirty-six hours before. People in traditional fishing ports who had always relied on ready supplies of fish for their own consumption found fish hard to get because it was all swept away to inland towns by rail. In 1865 fishermen at Prestonpans sent more fish to Manchester

Loading fish wagons at Grimsby.

than to Edinburgh, the latter only nine miles away. Before the railway age a large haul of fish often sold for a pittance because there was no way of getting it to market, and villagers thought supplies were inexhaustible. When the same people could no longer obtain fish they believed the shortage was caused by the use of the trawl.

In small fishing ports the morning's catch was laid out on the beach or quayside and a half-circle of buyers formed round the auctioneer who stood on a barrel or box, or at a kind of pulpit with a sloping tray on which samples of the catch were laid out. Some buyers were kept in touch with other markets by telegram, advising them prices were low because of an abundance of fish, or high because of storms affecting the Humber fleets. Buyers could have formed a ring that worked against fishermen, but competition was between buyers, not between buyers and fishermen. In the big ports such as Grimsby, crews returned to their vessels at about three in the morning to open up the hatches and dock labourers called bobbers swarmed aboard to swing the fish on deck by block and tackle and carry it under the roof of the fish market where it was graded. Cod was sorted into seven classes, the largest being laid out in neat rows. It was a 'Dutch' auction – the auctioneer sold at a furious speed, starting at a high price and moving rapidly downwards until a buyer raised his hand and shouted 'At!' which implied he would take some fish at that price. He then nominated how much of the fish he would take and his labels were placed on his choice of fish or boxes. Judging the condition of fish needs skill: the eyes should bulge slightly, the pupils be a glossy black, its skin colours bright, the flesh elastic and the gills (of gutted fish) a brilliant red. The truest test is to upset the container and see how far the fish slide – the further the better – but this is not a method guaranteed to make friends with fish salesmen.

As soon as the fish were sold there was pandemonium. In a welter of slush and slime the fish were divided, repacked, and rushed to the trains hissing in the sidings. The first London fish special left Grimsby at 5.20 p.m. and drew up at a special fish platform at Marylebone Station at 12.20 a.m. and a night staff unloaded it into drays. Other trains came in during the night, phased so that all consignments were in time for the 5.00 a.m. opening of what was known as the great fish emporium of the nation – Billingsgate.

The first Billingsgate dock, just below London Bridge on the north bank of the Thames, was probably built by the Romans. When Thames-side warehouses were re-built after the Great Fire they were set back to form a quay where goods from all over the world were unloaded from scores of merchant ships; Billingsgate dock was cut into this embankment and coal and fish were landed at its steps. It was always a lively, rough and tumble place ruled by its powerful body of porters and famous for colourful characters such as 'fish-woman' boxers

Auction of fish at Grimsby market.

who engaged in knuckle fights at public houses, each contestant holding coins in her fingers so that she could not resort to pulling the other's hair. From women such as these the word 'billingsgate' entered the English language as a synonym for foul and abusive speech. Billingsgate was declared a free fish market six days a week, with permission to sell mackerel on Sundays, by William III in 1699. Inspectors of the Fishmongers' Company were charged with the responsibility of ensuring that fish were fresh and fit for consumption.

Early in the nineteenth century the dock was filled in to make more room and scores of boats clustered round floating pontoons moored off the embankment to unload their fish. Before the railway age nine-tenths of Billingsgate's fish was delivered by water. In 1848 10,442 boats – peter boats with river fish, hatch boats with live cod from Gravesend, smacks with mackerel or herring – landed fish at Billingsgate. By the 1850s the amount of fish brought by water was equalled by that brought by land. The chief superintendent of London City Police estimated that the 1,900 fish carts calling at Billingsgate every day occupied an area eight times greater than the market itself. Everywhere is up-hill from the market, and the narrow lanes became hopelessly congested.

Until 1846 the market was in the open and salesmen and dealers protected themselves from summer storms and winter winds with large

green umbrellas or tumble-down shanties lit by spiked candles. A fountain in the centre was the water supply. Then a wooden roof was built and gas lighting installed; in 1876 the existing market building, double the size of the old one, was built. At that time there were very few fishmongers in the suburbs and men in the fish trade made a very comfortable living.

The early-morning scene at Billingsgate was one of the liveliest in London. Four or five steam cutters raced in from the North Sea fleets every night to be first alongside the 'dummy' or pontoon in the river. The twin gangways rang with the stamp of clogs and boots, and processions of white-smocked porters filed down one side and came up the other bearing boxes weighing as much as eight stones supported on hats padded with black leather; a scarf around the hard hat gathered the dripping slime. In the street one tide of drays bringing fish from the late trains met first waves trying to get away with con-signments to catch early trains for inland towns – just as today (though in smaller numbers) refrigerated lorries delivering fish from small ports all round the coast and as far away as north Scotland jostle with vans taking it away to retailers in the suburbs.

In a welter of slush and icy water fish were bought, sold and bought again, salesmen selling to 'bumarees' who broke the fish down into smaller parcels and in turn sold to West End fishmongers, hotel chefs, and costermongers. In the days of the well smacks the best-looking pair of cod was selected for first sale, then the best four, the best eight, and so on. The better the fishmonger, the earlier he arrived at the market for the pick of the day's landings. Haddocks were not even brought into the market until about seven o'clock, when the West End buyers had been and gone and the costermongers – four thousand of them in winter months – moved in with their hand-carts to buy cheap fish like haddock, plaice and sprats for smoking, frying, or hawking through the streets for sale in poor neighbourhoods. Coster-mongers took one third of all fish, buying haddock in ninety-pound lots for as little as a penny a pound. Most of it was sold fresh, but in 1881 there were at least 150 firms in London smoking haddock. Plaice was favoured by fish-friers who sold the fish from barrows.

A traditional sight at Billingsgate were the two or three Dutch eelboats, or scuyts, which had a prescriptive right to anchor in the river at least from the time of Charles I. When one left another always

---

13. *Fish auction at the harbourhead, probably in a Fife fishing port between the wars.*

14. *Congestion outside Billingsgate ("the great fish emporium") has always been a problem, which is perhaps the reason for the fish porter's reputation for bad language.*

took its place. The eels were kept alive in jar-shaped baskets in the hull of the boat which was pierced with holes so that the water flowed through it. About 800 tons of live eels were hawked through London streets, and smoked or jellied, every year.

III

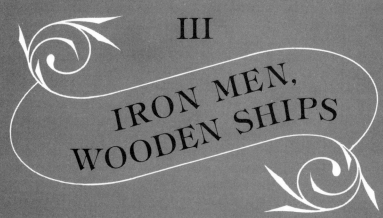

IRON MEN, WOODEN SHIPS

Water's hellishly wet.
(Fishermen's saying)

What Cheer, Ho?

The ordinary person buying a piece of fish a hundred years ago had no reason to suspect the almost intolerable industrial hell from which it had come, far out on the grey North Sea wilderness. By the 1880s working conditions in mines and factories had been mitigated by half a century of benevolent legislation; afloat, reforming influences had been at work in the merchant navy. But in the great fishing fleets scattered like depressed towns over the bleak sea, life was as close to slavery as anything that had existed in the dark days of industrialism. The public knew almost nothing of working conditions in the fleets, because until 1880 a death at sea in a fishing boat did not have to be reported unless the boat suffered damage.

From the 1850s until nearly the end of the century a community of up to twelve thousand men and boys was constantly afloat under sail in various fleets trawling the North Sea banks. Individual vessels joined the fleets and served seven or eight weeks of unremitting drudgery then made sail for a few days in port to rest and re-fit. But the fleets were always there, quartering the North Sea day and night, summer and winter, year after year. Men grew old and grey in them, knowing no other life until the almshouse claimed them in old age. The fleet was their mental and moral horizon for all their working lives. More than half the crews were boys, apprenticed from workhouse and orphanage from the age of about eleven, and paid only a few pennies a week. Absconders were marched to prison in chains (see page 114) but many boys preferred prison to the fleet because the Dogger in winter, they said, was worse than the gallows.

At their peak, during the 1880s, there were about fourteen main fleets and several smaller ones. The largest was Hewett's Short Blue Fleet of Gorleston, which had 170 trawlers crewed by 2,000 men and boys. Two other large trawler fleets fished from Yarmouth. Grimsby had two large and several small fleets that trawled near Heligoland in summer and the Dogger in winter; Hull had two large and three small fleets. The Ramsgate fleet trawled off the Dutch island of Texel in

91

summer and near the Suffolk coast in winter. There were small fleets at Scarborough and Boston. The Brixham fleet fished the English Channel in summer and joined the two Lowestoft fleets in winter. Leleu's, Columbia, Great Northern, Coffee Smith's, Gamecock, Red Cross – the great fleets were floating cities scattered on a waste of water.

When the smacks hauled their trawls at dawn they drew so close together that jibs and topsails seemed to form terraces of red and brown, some overlapping and seeming to touch, as if semi-detached, the ropes forming telegraph wires between them. As you sailed into the winding, changing streets there were large smacks sailing fast on every side, a dozen collisions seemed imminent but were avoided as the lumbering vessels ghosted up to each other and the smacksmen leaned over their bulwarks to shout the melodious greeting of the North Sea, 'What cheer, ho? What cheer?'

The fleet at sea.

On every heeling deck, wet with spray and gleaming with fishy slime, groups of fishermen stood in bulky brown oilskins, scarlet neck-ties, heel-to-hip leather boots, and weird assortments of head-gear speckled with silvery scales – caps of improperly cured sealskin that tended to send men prematurely bald, and stiff oilskin sou'westers as big as firemen's helmets and worn at every angle.

'Is that you Peter? What cheer?'

'A bad night Dick, only three trunks.'

'Ha! That's because you trawled up a dead body yesterday!'

Then the order of milling vessels changed as the 'suburbs' of the floating town interweaved and formed new patterns, and the gossiping continued with new sets of neighbours. Outwardly it was a lively scene of cheerful good-natured banter repeated in a hundred places every morning, but it was witnessed by few landsmen. The skippers were past-masters at their jobs but few could read and write, and the small number of logs that were kept in the fleets, and which have survived, provide only the barest details of fishing grounds and catches. No fishermen were literate enough, or willing, to describe the realities of their lives on the cheerless sea and it was not until missionaries began to visit the fleets in the 1880s that the true picture of life began to emerge. Only from their accounts is it possible to gain an impression of life in the fleets as it had existed during the previous thirty years.

One of the first landsmen to visit the fleets was a London Hospital surgeon, Frederick Treves, who recognised in the cry of greeting between smacksmen 'a spirit that was both manly and pathetic'. In its interpretation, he said, lay a sad secret of the sea, the whole secret of the fisherman's life in a North Sea fleet. 'So blank and dull is the life of these toilers on the deep that the one great thing they clamour for is a cheering break in their dreariness. The cry rings from smack to smack and in summer or winter, in fine weather or foul, it is an expression of the upmost feeling in the fisherman's heart – what cheer ho?'

At its best the living accommodation in a smack was pretty grim: a dog-hole of a cabin right aft, about eight feet long. Fishermen had barrel chests, and arms like derricks, but they were not tall (it was said that the tall species died out from falling overboard too easily) and there was barely five feet of standing room under the deck beams. Ventilation was nil because the hatch had to be shut to keep the water out and the tiny skylight was battened down to prevent the glass panes being smashed by seas, so the cabin was dark as a well. A steep wooden ladder gave access to the deck. On either side were two tiers of bunks, mere shelves with sliding doors like cupboards into which exhausted and sodden men crept, wearing full storm-kit except perhaps boots and sou'wester; once put on, their sea-going clothes were not removed until they returned to port. There were no sanitary facilities of any kind. Fishermen never washed, although areas of exposed skin saw plenty of water during the voyage, if no soap.

Half a dozen men and boys lived in this tiny space with a red-hot coke stove that leaked sulphurous fumes into the air in rough weather. Pipe smoke from very strong plug tobacco created such a fug you could hardly see one side of the cabin from the other. The fumes of cooking combined with the reek of damp oilskins and jerseys, the foulest stench from the continually agitated bilge water, and steam from socks hung from the deck-head to dry. Globules of tobacco juice spotted the floor and the dank air turned hinges and other brass fittings green with scaly verdigris. When the men came down from the deck and entered this stifling and poisonous atmosphere, where the temperature could be eighty degrees, the steam rose in clouds from their wet clothes and sweat rolled down their red faces.

As long as a man's stomach was strong he could at least survive. But in winter it was a different story. The little cabin was almost continually flooded, the fire was extinguished again and again by waves that followed men down the hatchway through which they jumped on the cry of 'Water!' to avoid being swept overboard. In clouds of steam the ashes and coals were distributed into the men's drenched bunks. During these long sodden days it was barely possible to rely on a hot drink, let alone a fry-up of fish and potatoes. Nothing more demoralised a smacksman than to be without his tea. He worked with a mug of it constantly at his elbow – a potent brew, stiff with body, which in the opinion of a visiting fisheries scientist in the early 1900s produced all the sensations of seasickness without the trouble of going to sea. It was made by boiling a large handful of tea and two pounds of sugar for some hours. Thereafter, water and tea were added when either appeared deficient, and the leaves were never emptied until the kettle was full. In 1891 several fishermen at Newlyn were confined to their houses after their condition was diagnosed as tea intoxication, or tea fever.

Smacksmen lived rough but seldom went hungry. If nothing else there was always plenty of fish. The crew had two main meals a day, a breakfast of fish in the middle of the morning after the night's catch had been taken to the cutter, and dinner of salt meat late in the evening after the trawl had been 'put to' for the night.

Cooking was done by an untrained, seasick and miserable youngster about twelve years old, and the food was often spoiled. Fish were fried or baked in a square 'kettle' like a high-sided baking dish with a lid, and vegetables were boiled in another. When the meal was cooked the kettles were placed on the deck of the cabin where the men's boots held them steady. Cutlery, if there was any, was stuck in the beams overhead. When fresh bread was exhausted the cook made 'busters', discs of unleavened bread marked with sinister shadows from his grubby fingers. One of the few women to visit the fleets at sea was Ada Haberson, a missionary, who remarked on the usefulness of a broom aboard a smack: 'I was much struck by its utility, especially in the washing of potatoes; the process may be useful to some on shore who do not know it. You fill a bucket with sea water, put in the potatoes, and stir well with the broom. Then you change the water, emptying the dirty water on the deck, and stir vigorously with the handle. Your potatoes are now ready.' Rats were got rid of in port by battening everything down but the forehatch, removing the ballast,

15. *Rare photograph taken aboard a trawling smack at sea as Skipper John Smith and his crew prepare to shoot their trawl in 1930; the horizontal taut rope is a relieving tackle on the tiller.*

then the crew advanced slowly from the stern towards the bows, making all the noise they could with fog-horn, tin-plates and spoons, and rocks rattled in buckets.

It was the procession of injured fishermen arriving at the London hospital from fish cutters docking at Billingsgate that prompted the surgeon Frederick Treves to assess the problem at first hand. 'Apart from the frequent exposure to cold and wet, the fisherman's life is dangerous. The aspect of a colony of some twelve thousand men unsupplied with medical aid should attract the attention of the charitable. If a man has a limb crushed, he may not improbably bleed to death; if this danger is over he still has before him a journey of two days in the steam carrier before a surgeon can be reached. Two days in a steamer probably seems long enough to the victim of seasickness: they must seem interminable to a man in acute pain, haunted with doubts as to the fate of his limb or his life.'

No stretchers were used, and the discomfort of being transferred from one pitching vessel to another by an open boat, with a broken leg or severe burns, can hardly be imagined. For the fisherman it was a case of work or go home. If you could not work you were a nuisance and if you went home you might not get work again for weeks. There was no compensation, no dole. Cures for sea boils and poisoned hands, which were common afflictions, were rudimentary. A poisoned hand was eased with a poultice of crushed biscuits in a wet rag. For boils and the deep cracks in the thick leathery pads of the men's hands, caused by intense cold and constant immersion in water, an ointment was made of sugar and soap, or paraffin and treacle. A cut was dressed with tobacco and tightly bandaged. Headaches, fever and stomach-aches were treated with a pill of Stockholm tar and flour. Constant chafing of stiff oilskin cuffs on unprotected wrists and forearms caused large excoriated and inflamed ulcers and boils which never healed as long as they were wet. During the voyage they collected a hard bracelet of scum and fish guts which were removed, when homeward bound, with a scrubbing brush and carbolic. After a season or two in the fleet few men could straighten out their fingers.

Danger, discomfort and hard work were hardly uncommon, ashore or afloat, a century ago. What made the smacksman's lot particularly hard and luckless was the unremitting monotony, the unvarying round of hauling and toiling in heat and cold, wet and dry, with nothing to brighten his life. He did not even have the privilege of the man doing the hardest and most dangerous job ashore, to sleep in his own bed at night. On his pitching horizon there was only one glimmer of comfort, one way of blotting out the discomfort and dreariness, and that was to signal for the grog ship. A skipper only had to hang an oilskin from an oar over the stern of his smack for the cruising 'coper' (see page 122) to round up into the wind and await his boat. Duty-free tobacco, at

1s. 6d. a pound instead of four shillings, was the main attraction, but it was hard to refuse the offer of 'one little drink' and by getting the fisherman to take a dram the coper sold a bottle. Many of the cheap spirits were rank poison. They made men incapable of looking after their vessels and many became demented. Often drink-crazed trawlermen bartered gear from their smacks, later reporting it broken or lost.

The men liked to play the accordion and some were good at it. Some made hook rugs for their families, or carved presents from hard salt-beef. Fishermen loved cats, but only black ones which were considered to be lucky, and some trained their cats to dive overboard for a swim. Dogs were often taken to sea, particularly curly-haired retrievers which fetched seabirds such as gannets, skuas and petrels brought down with fowling pieces. If not too badly marked by shot they were sold to bird dealers for stuffing and mounting, and decorating Victorian parlours. Just as modern trawlermen sometimes sport with gulls by blinding them in the beam of a searchlight and guiding them down and down until they touch the deck, lads in the sailing fleets caught gulls by towing a tarry rope astern in which a bird could become entangled, or knocked them down with osier wands as they swooped on fish towed astern. In winter the gulls could often be fed by hand.

Of the legions of small boys sent to the Dogger to perish or become heroes, many died or fled. But many also thrived on the rough sea life and overcame the initial shocks of unimaginable misery, hardship, and monotony, and served out their apprenticeships and became skippers. As an industry the fishing fleets were unique in that practically nowhere else was a poor lad with neither background nor education able to better himself so easily with nothing but the strength of his arms and the robustness of his constitution. Many became owners of vessels and some of whole fleets. Traders and merchants were only too willing to advance the necessary credit to a young man with strength of purpose. But the difference between profit and loss was a fine one and without additional resources a new smack-owner could be ruined by a winter storm or one unlucky voyage. Bankruptcies occurred frequently and many smack-owners found themselves shipping again as deckhands. One fishing-fleet admiral, described as ten times rougher and coarser than any of the other fishermen, had been brought up by a gentleman, was educated, and had been shorthand writer to the Bishop of Oxford. But he got into bad company, took to drink, and finally ran away to sea as a cook in a smack, working his way up until he was in charge of the whole fleet.

A missionary described another fishing smack skipper: 'A man so strongly built as to appear of average height, though really above it, carrying himself on a pair of shapely legs with an ease and grace borne of the constant need to keep in the perpendicular on the rolling smack, and a face which, as he says with a gravity and immobility of coun-

16. *North Sea weather: fishermen chat calmly as their smack dodges (lies hove-to) in what looks like a gale of Force 8 or 9.*

tenance characteristic of the sailor's joke, has for twenty-five years been the envy of all who know him. His dress is singular for, as someone says, his spars and canvas are all right, it's the hull that looks unsound. That is to say, from the waist upwards he is dressed in the ordinary guernsey and Tam-o-shanter but his nether attire is strikingly original. Having been reminded by sundry holes appearing that things do not last for ever he has cut off a portion of the legs of his trousers and patched them in such a way that the cloth trails after him, strongly resembling a sporran worn behind. But as primness of dress is an impossibility in the North Sea, little notice is taken of it beyond an occasional reference to vessels which need caulking. He enjoys a good joke, too, and spins a yarn well. There is a characteristic, pathetic humour in the tale he tells you down below over a friendly pot of tea. On a certain vessel (he says) the hands were below at dinner leaving a watch on deck composed of a man who stuttered badly and a boy who was bucketing down the deck. The man's face suddenly appeared, terrified, in the skylight.

' "What's the matter?" the crew cried.

' "The b-b-b-," stuttered the man, his fear making him totally unable to speak.

' "What?" they roared.

' "The b-b-b-," he stuttered, getting wilder than ever.

' "Sing it mon!" they shouted, knowing that a man who stutters can often sing clearly enough. From between his ashen lips he trolled out . . .

' "The b-boy's gone o-ver the side, b-bucket and all." '

When outward bound to join the fleet a smack towed a salted leg of mutton over the stern to wash out the salt and on reaching the fleet made a signal to invite a number of friendly skippers aboard for a 'muttoning party'. These 'cruising' (derived from 'carousing') days were the smacksmen's only social occasions, and were repeated when fresh meat and provisions were brought by the fish cutter. Unhappily practically none of the yarns survived the tellers but even those which have been recorded lose so much atmosphere. To appreciate a smacksman's yarn you have to imagine the weeks of work just completed and the weeks that lie ahead before you go home. You haven't washed for

---

17. *Hauling the trawl aboard was a back-breaking, dangerous task and fishermen were often carried overboard.*

18. *When the trawl was aboard the cod-end was hoisted up and the slip-knot released; bodies, fossils, old seaboots and trunks of prehistoric trees often spilled over the deck with the fish.*

19. *Gutting fish around the capstan of a trawling smack while the cook, usually a boy of twelve, selects the crew's dinner. Seaweed called "scruff" and boulders usually came up with the fish.*

weeks, your clothes are damp, mouldy and verminous. The crowded cabin is as thick as a London fog with steam and smoke, and as hot as a stoke-hold. The boy is on watch and astern half a dozen smack's boats belonging to the visitors bump together as the smack jills along. The deck lurches in the lumpy seas, there is a creak of timbers, a crack of canvas. In your fist is a mug of Dogger tea, as dark as stout and much more bitter. This is the yarn, you hear and – who knows? – it might well be true:

A smack's boy tried to commit suicide by jumping over the side but was saved by the skipper who feared the next attempt would be successful and decided to escort him ashore in a home-bound smack and hand him over to the police on a charge of self-destruction. While the skipper was away the mate, thinking he and his men should have some encouragement for working short-handed, told the decky to prepare a good dinner and make a plum duff in which he was to put at least four pounds of plums and a whole box of egg powder. This was an exceptional feast indeed and the boy could hardly believe his ears but he did what he was told. When the pudding was opened it had a rather peculiar odour and one deckhand remarked that it tasted queer, but the cook tucked into it and the others followed.

When they had stowed their portions under their guernseys one of the men asked whether the skipper hadn't put a box of rat poison on board. The cook admitted he had emptied the contents of an unlabelled box into the pudding. This had the effect of a bomb thrown into their midst. The men rushed on deck with burning throats. They threw themselves down and began to writhe in agony, wishing their pudding was anywhere but where it was. To attract attention they flew a flag at half-mast and it so happened that their skipper, returning to the fleet, was the first to see their distress signal. Quickly he put off to their help but as he sprang over the rail a pitiable sight met his gaze. His men lay about on the deck, moaning and groaning. Some had been sick. The mate raised himself on his elbow and said, 'We're all dead men, skipper!' Then he explained how the tragedy had occurred.

'Well that's a pretty do,' said the skipper, as he dived down into the cabin to see how much of his rat poison they had consumed. Next minute he was on deck again, shouting, 'That worn't no rat poison you put in the duff, that was my box of curry powder!'

ADMIRAL'S SIGNALS TO THE FISHING FLEET

(The admiral was a senior and experienced fisherman appointed to control the fishing movements of the fleet)

DAY

Flag at foremast head	SAILING (to new grounds)
Flag hauled down	TRAWLING
Flag at mizzen masthead	DO NOT BOARD FISH
Flags at both mastheads	SAILING BEFORE BOARDING
Flag at half-mast	CUTTER WANTED

NIGHT

White rockets at intervals	SAILING
Flare on quarter, and white rocket	TACK
3 flares and red rocket	TRAWL ON PORT TACK
3 flares and green rocket	TRAWL ON STARBOARD TACK
2 flares and two white rockets	HAUL
1 flare at masthead and one on quarter with white rocket	LAY-TO

If scattered by bad weather rendezvous at:

Tail end of Dogger	February 1 to March 1
Horn Reef lightship	March 1 to August 1
Clay Deeps	August 1 to October 1
Silver Pits	October 1 to February 1

Reef and Tops'l Breeze

The smacks that populated the North Sea trawling fleets were arguably the finest package of speed, power and seaworthiness ever seen in small sailing ships. Many a steamship rolling uncomfortably in a grey-green North Sea swell found herself overtaken by a close-hauled fishing smack, angling trimly up into the wind's eye like a witch. Such a vessel would have been the dream of mariners like Drake and Nelson, whose fighting ships, by comparison, had the windward ability of a cardboard box. The speed of smacks was deceptive because at twelve knots they whispered along with little fuss and in a smacksman's breeze of anything above Force 5 showed their heels to racing yachts. In calm conditions they were heavy and clumsy, with the ghosting qualities of a traction engine. But a smacksman didn't consider a breeze a breeze unless he had at least one reef in, and it had to be nearing storm force before he admitted there was 'a bit of wind about' and took in his tops'ls.

In the era of the clipper ship there might have been prettier and more elegant vessels than the fishing smack, but none was more successful in combining beauty with function, speed with power. You might stand at a harbour entrance on a calm morning and wonder as fishing smacks were towed clear of the land by tug and men and boys wrestled with the enormously heavy gear. The straight-grained trunk of a mast was as thick as a telegraph pole. Every timber in the vessel was massive, the best English oak. In the calm air the dark red and brown sails hung from the spars as stiffly as barn doors: to hoist the peak halliard of the mainsail took three men. But to the discerning eye of the real seaman there was something about the power and purposefulness of a fishing smack that instantly appealed: this was a vessel built to do a job.

Ability to keep the sea was the fishing smack's outstanding virtue and it took a real seaman to know the demands that were made on a boat by murderous winter storms. A smack did not begin to look interested in a sailing wind until it was strong enough to blow your hat

Sunday on the quayside at Lowestoft.

off; then the great sail bellied into powerful curves, the spars creaked as they settled into position, the mast bent a little under the strain, pulling the four windward shrouds iron-bar taut, and the heavy vessel, heeling a little, slipped forward through steepening seas with no sense of bustle. Plenty of ballast, the deep heel, and a long keel gave her an easy, comfortable motion in a seaway. For speed the smack had sweet, yacht-like lines, but not too much so. The bows were not so fine that they drove through the waves and made the boat 'slushy' to handle and wet to live in, nor were they so broad and buoyant that the vessel pitched heavily, throwing unnecessary strain on the the rigging. The stern was wide and flat to give buoyancy when the smack squatted stern-down in the water while towing the trawl; in the 1870s the square transom stern was superseded by the beautiful elliptical stern which was stronger but more difficult to build.

Until the 1860s most smacks were cutter rigged, but as boats became bigger the long and heavy boom was almost unmanageable and often dangerous. Many were then converted into 'dandy' or ketch rig which allowed the ideal balance of sail to be set in all conditions and boats became even bigger; this allowed larger trawls to be used and the old smacks which could no longer compete were sawn in half and lengthened by fifteen feet, being re-rigged with two masts. These typical new registrations at Grimsby show how smacks increased in size:

1850	*Friends*	39 feet (over deck)	25 tons
1860	*Olive*	59 feet	42 tons
1870	*Sunbeam*	71 feet	65 tons
1880	*Shells of the Ocean*	76 feet	80 tons

Smacks had no propulsive power but their sails, which were 'barked' to render the canvas more durable and prevent mildew; this was usually done by the skipper and crew themselves who once a year hired a barking yard and made up their own mixture of oak bark, tallow, yellow and red ochre and Stockholm tar which were brought from the copper in tubs and applied as hot as possible with mops. The resulting tints, from golden amber to deepest brown, depended on the age of the sail and the skipper's recipe. In dock the smacks were manoeuvred with long sweeps, which was excruciatingly hard work, but when steam-driven capstans were introduced in 1877 they could pull themselves about on very long mooring lines: the line was taken to a bollard far ahead by the smack's boy, a few turns were taken round the capstan drum, then the donkey engine was revved hard and the smack gathered momentum; then, dragging the heavy warp, the boy nimbly climbed round the outside of all the other boats lying against the quay and dropped the warp over the next strategically placed bollard.

A heavy tiller of carved oak curved up from the smack's rudder post to the height of a man's hips. The vessel's great sail power made it far from light on the helm and in anything of a breeze a relieving tackle had to be rigged between the weather bulwarks and the tiller because a man could not hold it unaided; frequently the man on the helm had to call for help. Because they were heavy and had such fine lines the smacks took a lot of stopping, and when sailing into harbour sail was progressively reduced and the heavy bowsprit, which thrust fifteen feet outboard, was run inboard on rollers. Just in case an emergency stop was required a heavy canvas drogue was put ready by the stern; it could be thrown overboard and act as a brake, like an aircraft's landing parachute.

Fishing smacks were the last large sailing vessels to be built by eye, incorporating all the wisdom of shipwrights and the sea experience of fishermen which had been handed down through generations. Every curve was a natural curve of grown timber hauled from oak forests by 'tugs' of heavy horses. Timber was seasoned at least a year for every inch of its thickness. Smacks were overcome by breaking seas on the tail of the Dogger, run ashore when skippers misjudged narrow entrances like those of Great Yarmouth and Lowestoft, and were blown up when crews deliberately jammed the safety valves of donkey-engine boilers to obtain more power in the capstan. But fishing smacks never fell apart. Long after steam had made them redundant they continued fishing under foreign flags. Some ferried fish from Iceland to England during the Second World War, and worked in Faroes waters up to the 1950s.

That a smack would have to face weather conditions as bad as anything the sea was capable of was a certainty, and it would happen not once but many times in a smack's life. In anything less than a stiff wind of fifteen to twenty knots the smack did not have enough power for trawling. The ideal trawling wind was one that required a single reef in the mainsail but was not so strong that the topsails had to be handed. To reef, the cook was flung bodily into the belly of the mainsail which was often full of water when the smack was hard-pressed and heeling over. Then he worked his way along the sail threading the lacing through the eyelets. If the sail flapped at this time (which was unlikely to happen, but was not unknown) the boy was catapulted into the sea and inevitably lost.

It was blowing a snorter (about Force 8) if the topsails had to be struck and a second reef hauled down in the mainsail and mizzen. A canvas dodger was then erected in the rigging to protect the helmsman. The decks would be awash but the heavy and deep smacks rode this kind of weather easily and it was unusual for fishing to be discontinued. The weight of the trawl towing over the stern gave the vessel steadiness. The dangerous time came when the trawl was hauled in, and the eight-inch manilla warp was liable to snap as the gear neared the surface. If third and fourth reefs were needed the weather was severe indeed. The skipper would now heave-to on the port tack so the two-ton weight of the beam was on the windward side, then he went below and steamed his feet in front of the fire, stirring only to call out to the man on watch if a wave swept the deck.

The trawl beam was two or three feet in circumference and made of two lengths of Belgian elm with the butts scarfed together. It was heavy because the wood became waterlogged. One of the two-hundred-weight wrought-iron trawl heads slotted through a gangway in the bulwarks near the main shrouds, and the other hung outboard level with the counter. The length of the beam (and therefore the size of the trawl) depended on the length of the smack. The biggest was about fifty-two feet, but beams of forty-eight feet were more common in the typical seventy-two-foot trawling smacks of the 1880s.

To shoot the trawl the wind was spilled from the sails until the smack was barely moving; then the net was paid over the side and allowed to stream away to windward. The heavy beam was unlashed and the forward trawl head was levered over the side until the whole beam dropped into the sea, being prevented from sinking immediately by the trawl warp which was tight on the 'dummy' or towing post. As the sheets were hardened in to get the smack moving the warp was slowly paid out, the forward end of the beam swinging outwards and the aft end being held close to the stern of the smack by a line called the dandy. As the beam came square on the dandy was let go and fixed to the trawl warp with a clove hitch where it would be ready for use when hauling the trawl aboard again. With the smack gathering way the warp was allowed to slip round the dummy so the beam was slowly pulled along as it sank, ensuring it did not land upside down. In calm conditions less warp was paid out for the depth of water so the beam was lifted slightly to reduce friction; in strong winds more warp was paid out so the trawl did not travel too fast.

When the required length was out a 'stopper' of thinner rope fixed to the dummy was made fast to the warp so it took the whole of the strain of the trawl. A bight of the warp was then carried forward, outside the shrouds, and led inboard through the bow fairlead where it was securely fixed to the windlass. If the trawl snagged an underwater obstruction such as a wreck the stopper parted, the bight of the warp flew outboard, and the whole of the strain came on the bows which automatically brought the smack up into the wind, stopping her dead, and prevented the main warp from parting. While towing, the trawl warp was led from the dummy outboard over the rail. In the rail were holes in which thole pins, like large wooden pegs, could be placed. By altering the position of the warp with thole pins the angle of the smack's head was adjusted and with trawl down and a fair tide the vessel could be made almost to steer herself. The drag of the trawl was very great, reducing speed from eight knots to one knot. For this reason the sailing smacks nearly always trawled in the same direction as the tide, so it helped them along.

To haul the trawl the smack came into the wind and the warp was brought in through the gangway in the bulwarks. Before the steam capstan, heaving on the capstan bars or sweating on the winch handles was fearsomely hard work. It took three quarters of an hour at the best of times, but three hours or more in bad weather. As the warp came slowly inboard the cook stowed it in the warp room beneath the mainmast. When the dandy line came into view it was unhitched from the warp and taken aft where in some vessels there was a small winch called a dandy-wink. First the aft end of the beam was hauled up to the rail by the dandy line, and securely lashed down, then a hook on a tackle attached high on the mast was used to haul the forward trawl head aboard. The net was gathered in by hand until the cod end was floating alongside. Then a strop was put round it and the bag of fish was hoisted over the rail. The line tying up the cod end was pulled and the fish poured on to the deck.

In the ruck of fish would be clumps of grassy seaweed, sea-slugs, starfish and other creatures which the fishermen called brash and immediately threw overboard. There were also boulders which caused great damage to nets. Robert Hewett had his Short Blue fleet clean up the North Sea by bringing the boulders home and landing them on Gorleston Quay; later he used them to build the foundations of a mansion, which is still standing, and also the cottage hospital. From the Dogger bank ancient tree-trunks and the bones of prehistoric animals were dredged up, relics of the days when it was a plateau above sea-level. Pieces of amber found in the trawl were polished up and made into jewellery, and many smacksmen carried pieces in their pockets to ward off the fishermen's scourge, rheumatism. For the smack's mate tugging the line to open the cod end there were often some grisly surprises: in the cascade of fish and brash slithering around his knees might be the dead body of some fisherman. Usually the body was searched then thrown overboard again, and if any of the clothes or boots were still in good condition they were removed. If the body was not badly decomposed it might be put aboard the steam cutter in case it was recognised by any of the men boarding fish. In 1911 George Mooney, deckhand of the steam trawler *Uxbridge*, was carried over the side while the gear was being shot – and fished up again in the same trawl by the same ship on the following day. One sailing smack fished up three dead men in one haul.

Spending about forty-two weeks a year with the fleet, not including time spent sailing backwards and forwards to port, and allowing for time lost due to gales and calms, a Grimsby trawler like the *Angelus* in 1867 made 274 hauls of fish totalling ninety-six tons. In the early days of North Sea trawling it was not uncommon for a smack to take two or three tons of fish in a single haul, but the average was lower. In the most general terms a fishing skipper earned about one penny an hour, twenty-four hours a day. In Grimsby this was made up from a wage of about fourteen shillings a week plus one shilling in the pound of the smack's earnings. After a two-month voyage a skipper was well pleased if his catch made £150 and he drew £7 10s. for himself, but £5 was nearer the average. The mate was paid 18s. a week, the third hand 16s., the fourth hand 14s., the deckie 12s. and the cook 10s.; apprentices, however, were paid much less. The men also earned a little extra pocket money by saving the small, unsaleable fish that were caught towards the end of the trip and selling to a local fish-frier who met incoming smacks and offered two or three pounds for the lot; these were known as stocker, or in Scarborough as wrangems. Alternatively, the men might be paid shares of the smack's catch, in which case they paid for their own provisions. In a Grimsby trawler in 1877 the total receipts after the catch was sold were divided into eight shares, of which the skipper took one and three-eighths, the mate one and one-eighth, the third hand the same (or £1 a week and his keep) and the owner four and three-eighths. In its day it was not a bad living. A fisherman could go into a bar, call for a pint of beer and an ounce of shag, and get a penny change from sixpence. A glass of ale and a meat sandwich cost 4d., brandy was 2s. 8d. a pint with 2d. back on the bottle and a suit could be made to measure for thirty shillings. But to achieve this modest standard of living the fleet fisherman, like the trawlerman today, had to work at least twice as many hours as most landsmen.

Smacksmen grew up on the North Sea and learned its geography by touch until they knew it a great deal better than their own beds at home, in which they slept for only about eight weeks a year. Equipped with an apology for a compass, which was seldom adjusted and which was liable to freeze solid in winter, and only a lead-line for measuring depth and a log for measuring distance run, skippers located their fleets more than two hundred miles offshore, fished for two months, then found their way unerringly homeward. Not even the admiral of the fleet carried a chronometer which enabled him to obtain longitude, and anyway he would not have known how to use it. Navigational aids were almost non-existent except for the forty lightships or so in the English Channel and North Sea. A smack homeward bound from the fleet would first look for a particular lightship from which the skipper set a course for home without reference to a chart. So well did the smack skippers know the North Sea that when hailed by a merchant-man they could instantly give the bearing and distance of the nearest lightship without having to think about it.

Before 1880 any man who had the owner's confidence could take a smack to sea as skipper, and there were cases of vessels putting out under the nominal command of boys only twelve years old: they were sons of the owners, and therefore could keep the highest share of the operating profit in the family. The real skipper was the mate, who was sweetened with a shilling or two extra. Every year each smack was surveyed by directors of the local insurance club, themselves generally retired skippers or owners. In 1880 insurance clubs also began examining skippers and mates for professional competence and issuing certificates; an official certificate of competence issued by the Board of Trade was not required until 1883. These were very easy to obtain, as the exam comprised only a few questions about the rule of the road and simple dead reckoning navigation, and were put to the candidate orally so he did not have to read or write. Gradually standards were tightened up, but many skippers who had never sat an examination and could not read or write were able to continue going to sea until well after the First World War.

THE LIFE OF A SMACKSMAN

(A conversation with a Great Yarmouth smacksman recorded in 1909 by the Norfolk naturalist A. H. Patterson.)

Look here, I've been all through it; it's a rum, hard life; I'd as soon be a conwick as go through it all again. Things wasn't then like they are now; you lived on rough tack, and it grow'd a bit monotonous. We'd boiled fish for breakfast – which was mostly offal – dabs, whitings, garnets – we dussent tackle the prime, not as a rule, anyway, for turbots and soles were not for *us*. For dinner we got salt meat two or three times a week, and not the tenderest or youngest at that, 'specially the pork! We had brown biscuits – made of bad flour no doubt – they called 'em smack's biscuits. One skipper I was with wouldn't let us soak 'em in our coffee, nor yet split 'em, and we had only a thin smear of butter to help 'em down. We didn't often fry our fish, though we did chance times – using oil from the cod's livers to cook 'em in. Two or three times a week we called 'branyan day' – they was boiled rice and treacle days.

I'm speaking of fifty year ago. We used to start fishing twenty miles east of Lowestoft, then, when fish got a bit scarce we went to Botany Gut, off the Humber. We used to get as many fish then in a six hours tide as they now get in six weeks. The steam trawlers ripped up all the 'ross' ground and killed the feed. In all weathers they're rippin' and tearin' about, and nothin', neither fish nor feed, get half a chance.

We used to take in about two ton of ice afore we put to sea; and icing was a regular payin' business in them days. There used to be three great thatched ice houses always busy; a fleet of wherries used to bring from twenty to thirty tons apiece from the rivers and broads to the ice houses, when they were froze out from navigation, and it paid 'em well. Then the Norway ice came over by ship-loads and cut them out, as ice-factories have ousted *them*. Things fare always changing about, and for the workin' men – for the wuss.

The cutters came every day, when they could find the fleet; sometimes they'd miss us, and I've known a night's takings thrown overboard when the ice was short and the cutters hadn't come. The third and fourth hand took the fish in the small boat to the cutter; I've seen men and boats swamped, and go down afore help could get to 'em. I shall never forget one trip, it was blowin' a fearful gale of wind, and the fourth hand refused to go to the cutter; we'd got a nice lot of fish, and it was near the end of Lent, so I volunteered to go, bein' wrong for a mate to do so. We got to wind'ard of the cutter, and over with a can of oil, and while it was workin' towards her, slipped the little boat in and went along the smoother water. As it was, my hair stood on end, and I wasn't generally nervous. I don't wonder, with all the men lost around the Dogger, we sometimes got human skulls, and that, in the trawl, and more 'an once I've seen a man in oily jumpers, and boots on, shot out on deck with the fish when we've opened the cod end. It made your flesh crawl, shovin' of 'em over again with a capstan bar or your shovels. Lor, the Dogger's a regular ocean cemetery. . . .

Smacksmen won't be despised; they might be ignorant of book larnin', and perhaps the only one able to keep accounts and write out the bills of ladin' was the cabin-boy, but they was no duffers at sea. If they didn't make much out of the chart they got to know to an inch where they wor – and a lot better than the chart could figure it out for

111

them; they'd grope their way about by the lead by day or night, and when a scientific skipper would be fumblin' about in a fog they'd know their bearin's to a hair's breadth. And then for pluck, they didn't know what danger was; there was heroes in the fleets what oughter had medals and pensions, but never a word of it got wind except some newspaper bloke out figgering up some terrible gale and reglar pumped information out of 'em. You know as well as I do how many a vessel in direst peril would have gone altogether, crews an' all, but for assistance rendered by smacksmen at the risk of their own lives. People ought to think well of the smacksman, for his hardihood, let alone for the price paid for the fish they eat, what have often been caught for them at the sacrifice of a fisherman's life.

Sentenced to The Dogger.

Thousands of miserable little boys were sent to North Sea fishing fleets to perish or become men and heroes. One in fourteen died or was killed in the attempt. Scores suffered cruelly at the hands of men who themselves had grown up during the merciless era of the lash. The smack's boy was at the beck and call of every man on board. On a pitching deck far out at sea he had no appeal, nowhere to run except the route known as the fisherman's walk – three steps and overboard.

For every skipper who did not speak to his boys unless it was with a rope's end, there were scores who did have some regard for their welfare. But in the stern Victorian times a man who displayed a sense of responsibility was not necessarily kind hearted, and many of the most tyrannical acts were committed by crew members only a little older than the victims. A boy learned his duties the hard way. One skipper who was apprenticed in the 1860s said that when he was cook the skipper came and looked at all pots, pans, knives and forks and cupboards every night to see if things were in order. 'I happened to leave a spoon out one night. The skipper waited till I got asleep then called me out to find what was missing. I overhauled everything and at last found I was a spoon short, so I had to find it before I could go to bed again. It took me an hour and a half to find it – stuck into a rope at the masthead, so I never left things out of place any more after that. I think I got more hidings than good dinners, and the fishermen as a class of men were cruel to the boy apprentices.'

A boy's work in a smack was never done. He cooked meals, cleaned the cabin, attended the men. On deck he kept the gear tidy, scrubbed away the fish guts and scales, and looked after the lights which had to be filled with paraffin, trimmed and their lenses cleaned. While the trawl was shot he steered the smack. When it was hauled he went into the warp room, a dark hole at the foot of the mainmast, to coil the warp as it came off the capstan. In freezing weather this was anything but pleasant because he had to stand with his arms lifted up, clinging with all his weight to the heavy sodden warp as water ran down his arms, wetting him through. If the rope slipped, which happened frequently in rough weather and was not always his fault, a bucket of water was thrown at his head, the bucket being kept close to the hatchway for the purpose. His sleep was broken by constant demands of the crew to fetch things, run messages to the man on watch, and make tea.

The apprenticeship system had grown up with the fleets and was essential to their operation, particularly in Grimsby where population and labour supply lagged far behind the growth of the port. The 625 smacks registered at Grimsby in 1878 were crewed by 1,680 men – and 1,800 boys. Fishing did offer a poor boy one of the best chances he could find for success in life, and about half the smack-owners in Grimsby had themselves been apprenticed from workhouses, unions, reformatories and orphanages. Boys from public institutions were said to be the best lads in the trade and comprised about forty per cent of the total. But the system was (by modern standards) outrageously abused. Only one boy in five had permission of his parents, one in ten was the son of a man in the fishing trade, and thirty per cent were tramps. Smack-owners as a body took apprentices that no other trade would look at. Boys arrived at the port shoeless and in rags, having tramped across from industrial cities such as Leeds and Sheffield to seek their fortunes. A boy of eleven could find himself bound for ten years without realising what he had done. Indentures were drawn up (for a small fee) by a pensioner from the Customs who shared an ante-room with messengers in the Customs office but in fact had no connection whatever with the officials. A parent or guardian did not have to be a party to the agreement, its conditions were not made plain to the boy, he was not supplied with a copy, there was no system of payment of adequate wages when he became old enough to be really useful, and if he broke the agreement he was gaoled.

Under the Merchant Shipping Act 1854 the master, mate, owner, or ship's husband of a fishing smack, or the police, had the power to apprehend any deserter without warrant and take him back on board or to court where magistrates were entitled to sentence him to twelve weeks' imprisonment for desertion or ten weeks for disorderly conduct. Hundreds of boys were sentenced to hard labour for refusing to sail or absconding, often leaping ashore and running off just as the vessel began to move away from the quayside. About a third of those apprehended were not prosecuted but sent to sea again when the skipper or owner interceded, but it is not clear whether this happened because they did not want to see the boys imprisoned, or because they needed the boys' services. In 1877 at Grimsby 461 boys were charged and 200 released to join their smacks. Of the 261 sent to court all but three were imprisoned, 51 for the second or third time; all were under twenty-one and some were only thirteen years old. In the first six months of 1878 no less than 229 boys were imprisoned, the average sentence being three weeks for first offenders, longer for second offenders. The Governor of Lindsey Gaol told a House of Commons committee: 'The place is warm and well ventilated so they don't mind going to prison during bad weather in winter – of course they don't get as good food as outside.' One Grimsby boy, Edward Tevereux, was told by the Bench he could go back to his ship if he wished but after consideration he said he would rather go to prison, and was sentenced to fourteen days with hard labour. One boy from a Nottingham workhouse finally had his indentures cancelled after three years during which he had been imprisoned five times for a total of five months. A similar situation existed in Hull, but sentences were, if anything, harsher.

In 1873 two articles in the French newspaper *Le Figaro* alleged that at Grimsby there was a system of slavery as infamous as any ever devised and this was widely reported in England. The *Lincolnshire Chronicle* was outraged. 'The city of Lincoln rings with indignation at the treatment these lads receive at the hands of the authorities. The lads are brought (from Grimsby) by train, which generally arrives at 9.50 p.m., and heavily chained together in numbers of three to five are marched through the busiest part of the High Street for more than a mile to the gaol.' The Holborn Board of Guardians, which had sent many orphans and poor boys to be apprenticed, called a meeting and sent its secretary to investigate these allegations. He reported that the chain was only a light one, to which four or five handcuffs were attached so one policeman could take several prisoners 'and avoid unnecessary expenditure of public money.'

The demand for apprentices at Grimsby led to what the *Leicester Mercury* in February 1878 called legalised kidnapping. The newspaper described how young 'decoy ducks' were sent out to lure teenagers to Grimsby with romantic tales of a life on the rolling deep – 'where he is too often doomed to commence a life of hardship, wretchedness and brutal slavery. The parents are simply ignored. The very infant who in the eye of the law is considered not responsible for the slightest debt is entitled to be beguiled, cozened and cheated into self-surrender to a white slavery as cruel as it is degrading, while his hapless parent is utterly powerless. An Englishman's house is his castle but his son is at the mercy of any unscrupulous crimp.'

Hundreds of apprentices broke loose in 1880 when the Payment of Wages Act abolished the power to arrest deserters without a warrant, causing chaos in the manning of the fleets, particularly in Hull. Under the Merchant Shipping Act of the same year, wilful disobedience of orders was still punishable by a prison sentence. In Hull this was interpreted to apply only at sea, but in Grimsby smack-owners claimed it applied to apprentices ashore or afloat. Meanwhile, partly in response to two lurid murder trials for the killing of apprentices in fishing smacks, Grimsby owners voluntarily overhauled their system. A Board of Trade committee supported the improvements and by 1893 they were legally enforceable. Under the new rules no boy under the age of sixteen could go to sea unless he was an apprentice, and no apprentice could be younger than thirteen. Boys could spend a trial period of up to six months before they signed indentures under the direct supervision of a Board of Trade officer who explained to them their rights and responsibilities, but, once he signed, an apprentice was still liable to be imprisoned for desertion or wilful disobedience of orders at sea or while preparing for sea.

There were two types of apprentice at Grimsby. The 'indoor' apprentice was clothed and provided with accommodation, often in the smack-owner's home, and given a few pennies a week as pocket money. The 'outdoor' apprentice was paid seven to sixteen shillings a week, out of which he had to buy his own clothes and fend for himself when the smack was in port. He had no supervision and mixed with casual hands of the same age but who had a lot more money. Owners unable to lodge indoor apprentices in their own homes boarded them out. Respectable families willing to receive scruffy fisher-boys and take an interest in their welfare were hard to find, and some boys were not really fit to live in homes that had standards of behaviour, dress and cleanliness. Some boys were lodged in known brothels but in the main they were reasonably looked after and the secretary of the Holborn Guardians was impressed with the way boys were kindly treated and 'mothered' by smack-owners' wives. Clubs were started to attract boys away from brothels and bars and provide what was described as moral, spiritual, mental and social welfare. A Fisherlads' Institute was opened in 1879 and used by 10,000 boys a year. This greatly reduced the number of charges of disorderly conduct against fishing apprentices, which in 1877 had totalled 461 in Grimsby and 320 in Hull.

Boys who were not sufficiently robust suffered terribly. In 1893 the master of a Grimsby cod smack was fined £2 10s. for a breach of the new act by taking a boy aged fourteen to sea without giving him the opportunity to visit the Mercantile Marine Office. Reports said 'the poor boy was continuously sick the whole time and died from exhaustion at sea, being obviously unfit for his work.' Before the Act, however, this kind of treatment was routine and a death at sea did not even have to be reported unless damage to the vessel was also involved.

This was the experience of one North Sea smacksman who was apprenticed in 1873 at the age of eleven: 'While sitting at my copy-book in school our master called out, "Is there any boy wishes to go to sea?" With four others I held up my hand and the smack-owner who had come picked me out. I went with him to Ramsgate and had been there only three or four days when I tasted salt water for the first time, having fallen overboard in harbour. The first week I didn't know much, as I was sick all the trip, but on the next trip I had to do my work, cooking, trimming the lamps, and so on, and at everybody's call. I can remember us pulling up a cask of wine. The skipper and mate got drunk and because I couldn't pull the warp back when we were hauling they threw pails of water over me. Another time, because I told our skipper a lie, he knocked me about with a scrub-stick until it broke, and then he made me keep on deck for three hours, carrying a heavy handspike. Every time I came aft I had to say "All's well". In summer we fished twelve weeks with the Lowestoft fleets and our skipper would go to town by cutter to get fresh food. I remember once, while frying fish, I laughed when the skipper told me to pump the smack out, as she made a lot of water. I said to the cook, "I expect she's full again." The skipper heard me, knocked me down the hold, then threw a fish-trunk at my head and I have the scar to this day. We were always wet through and if we asked our owner for some clothes he stopped our spending money. The decky had sixpence and the cook threepence a night when we were at home. The owner would say, "You'll have clothes when I know you want them." I served seven years and very glad I was when my time was up.'

The murder trials which so raised public indignation over the plight of boys at sea brought to light particularly lurid details of cruel treatment and showed how boys in fishing smacks were completely at the mercy of older men in the crew. Aboard the Hull smack *Rising Sun* the cook William Papper was thrashed repeatedly with rope, hanged from the lantern halyards and knocked unconscious when the cross-trees broke and he fell to the deck, and in December weather was made to stay on deck four days and nights without a change of clothes. Skipper Osmond Brand and third hand Fred Ryecroft lashed the boy naked to the rail and threw buckets of icy water over him, made him crawl into the dill (a hole in the floor of the hold giving access to the bilges, which were half full of water) and jumped on him to make him fit, struck him with baulks of wood and made him stand on the taffrail singing songs while stones dredged up by the trawl were thrown at him. For Christmas dinner the crew had two ducks and plum pudding but Papper was given only the bones after the dog had picked them. When the boy died in his bunk the skipper dressed the body and threw it overboard and reported he had fallen over the side. Brand was later hanged at Leeds.

Edward Wheatfell, mate of the Hull smack *Gleaner*, was executed for murdering the Grimsby lad Peter Hughes who was thrashed repeatedly, stripped naked and made to carry a bucket of water on his head in keenly freezing weather then forced to stay on deck night and day. The boy disappeared while only the two of them were on deck, and Wheatfell claimed he had fallen overboard while drawing water with a bucket but was later convicted of murder.

For a young boy the terrors of the sea alone were frightening enough without the infliction of hardship and cruelty. The Brixham smack *Ruby* was towed into Hull with only a very small boy, John Lill aged fourteen, on board. He had a hazy recollection of a crash as a German barque collided with the smack, and seeing the smack's crew jumping aboard the other ship, then he was alone. He stayed on deck all night pumping and crying until a trawler bore down on the wreck and saw him waving his cap with one hand and pumping with the other. Another who found himself in a similar predicament was Charles Taylor, also aged fourteen, who told his story to a newspaper in 1891: 'Last Tuesday night I was at sea fishing in the *Silvery Dart* when the wind came on to blow very hard. At about ten o'clock I was asleep in the cabin when a vessel struck our boat forward and I ran on deck to see all our crew jumping on board. I called out and Thomas Overbury shouted to me to show a flare. They left me alone on the vessel, thinking she was sinking. We had only one flare, which burned for about an hour then went out. I stayed on deck until morning, and all next day, and next night. I was too frightened to eat. Early the following morning I ran foul of a Dutch herring boat drifting with her nets out and the skipper told me to let go my anchor but I was all alone and couldn't do it.

'All that day and the next night passed without my eating anything. At twilight next day some smacks saw me drifting but didn't come close enough. On Friday morning I saw another Dutchman and waved my sou'wester as he ran down to me. I jumped aboard as the other boat sheered close to the port quarter and the crew caught me. We left the *Silvery Dart* behind us and the Dutchman kept on fishing until the end of the week. They treated me kindly, giving me clean and dry clothes and plenty of food. In port I went home with the mate to sleep and next day went to the skipper's house for dinner. He put me in a train for Rotterdam and I reached Yarmouth on Sunday night to find that my family had gone into mourning for me because they believed I had been washed overboard and drowned.'

Devil's Work.

After a night of trawling the crew of each smack gutted the catch and sorted it into boxes. The day would be spent tacking up-wind to make good the leeway lost during the night, but first, in the early hours of daylight, came the job smacksmen most dreaded – boarding fish. It was so dangerous that lower ratings often resisted promotion to third hand and mate to avoid the task of rowing the fish to the steam cutter in the smack's boat. Few smacksmen could swim and their clothes weighed twenty-five pounds. When their heavy leather seaboots filled with water they went straight down. And iron nerve, giant strength and wondrous skill were no match for the mischief of a curling wave that knocked a boat, top-heavy with its load of fish boxes, head over heels. 'Boat upset while boarding fish' was the epitaph of many fine smacksmen. One fleet lost thirty-seven men while boarding fish in a year.

The steam cutters were fast, racy vessels about 130 feet long and 22 feet wide, with high cold-chisel bows sloping down to a graceful counter stern so there was a low freeboard amidships. Their bottoms were reinforced against pitching, for they travelled at full speed in all conditions, barrel-rolling through hard head-seas with spray driving like concrete chips high over the meagre canvas dodger sheltering the helmsman on the bridge. Finding the fleet was often difficult, particularly in fog when the admiral of the fleet fixed gun rockets to the rail and fired them with a red-hot poker to keep his boats together. On finding the fleet the cutter first handed out empty fish boxes to the smacks which sent their boats for them, then might use its own trawl to make one or two hauls before its turn came to hoist a flag and announce that it was ready to begin loading fish.

Aboard each smack the heavily built boat, about sixteen feet long and seven feet wide, was lifted stern first on to the lee rail and slid outboard until it reached the point of balance. Then the painter was made fast and as the smack rolled to leeward the boat was launched with a hard shove. The boxes of fish were handed down to the men in the boat, sometimes forming a stack four or five feet high amidships. The men rowed in bow and stern, standing to their oars so they could see oncoming waves and also converse.

Meanwhile the smack had sailed to windward of the cutter and when it was conveniently close the boat was cast adrift and laboriously rowed around the steamer's bow (never the stern, which overhung the water like a yacht's counter and was liable to 'sit' on a boat that passed beneath it). The smack then sailed round to await the return of the boat to leeward of the cutter.

In a calm sea with nearly 130 smacks launching small boats carrying nearly three thousand six-stone boxes or trunks of fish, which converged on the rolling cutter with its narrow Woodbine funnel and tattered boarding flag, the scene was a merry one. There was a good deal of hearty recognition among men who knew each other, and some robust language as gunwales bumped, oars were fouled, and fishermen with painters in their teeth leapt for the cutter's low rail. A 'fish note' giving details of the catch was dropped in a basket in the cutter's galley, and the boxes of fish were passed up over the rail to be stowed away in the hold and sprinkled with crushed ice by the boat crews of two or three smacks who were hired by the cutter's master to assist his own men. The cutters were postmen for the fleets and occasionally a smacksman found a letter from his owner, or from a sweetheart, or a parcel of fresh provisions.

In bad weather the task was hazardous in the extreme. When the fish had been caught and packed ready in boxes to be taken to the cutter it was unusual for a fishing skipper to tell his men not to risk it. They would have seen boats launched in worse weather. They would be unwilling to lose the fruits of a night's toil, especially as the best prices were obtained when supplies were short due to bad weather. Loading the boat from the smack was difficult enough, but the smack was steadied by its sails. The cutter rolled like a thing demented and with a press of small boats around it, all heavily loaded with fish, the dangers and chaos can be imagined. Once he got alongside, the fisherman in the boat balanced himself lightly and lifted a box of fish to waist height. Then, in the split second that the cutter's rail dipped low towards him, and paused before beginning its upward travel, he laid the box of fish half across it. The man on deck steadied the box as the fisherman let go. As the rail soared upwards the box slid off it and the man on deck guided it to a soft landing. The skill and agility of this work was the admiration of landsmen who visited the fleets in the cutters and themselves faced the daunting challenge of reaching a smack by one of the boats. Dr Wilfred Grenfell, surgeon and missionary who once saw fourteen boats upset in one morning, had his own recipe for getting on board a boat:

'Lie full length on the rail, taking care not to get your arms and legs jammed.

Wait until the boat bounces in somewhere below you.
Let go.'

Often the small boats were cracked like nuts when the cutter rolled on top of them. Conditions were judged to be too bad for boarding fish when the cutter rolled each rail under. A writer of the *Yarmouth Independent* described the wildness of the scene when he visited a fleet in a cutter one January: 'The wind continued to increase, and by the time half the fleet had delivered it was blowing half a gale. The main-decks were piled up with boxes of fish and still there were several boats watching for their opportunity to get their fish out. The steamer rolled so heavily as to twice fill my sea-boots while standing on the quarter-deck so, clearing out of that, I got on the bridge and had a good view of what was going on. She had now shipped several rather heavy seas, some of which found their way over the high coamings and down the main hatchway. Just now a boat, with several boxes of fish and two men in it, was landed high on the steamer's rail and on the pile of boxes, and if two men who were close hadn't pushed her off again she would have been lifted high out of the water and turned a complete somersault.'

When two hands were lost from the Hull smack *Abo* when their boat was upset while returning from the cutter *Northward* on August 9, 1887 it was decided to hold an inquiry at Hull before a magistrate and two sea captains. Between that date and November 10 a further ten lives were lost from Hull vessels, both smacks and steam trawlers, while boarding fish. The Fisheries Inspector made a special report on the matter, suggesting improvements such as a supply of lifejackets and lifebuoys. In 1894 uniform regulations were established. The admiral was to decide when it was safe to board fish, and if the press of boats alongside became too large for safety the cutter would lower its flag to half mast. Trawlers had to be provided with two lifebuoys and three lifejackets, and boats were fitted with a stern-ring to which lifelines were attached underneath the boat so that a fisherman had something to grip if it overturned. But fishermen were often reckless and always afraid of scorn, and the cumbersome cork lifejackets were never worn.

It was after boarding his fish that the fisherman was liable to fall victim to the coper, or grog-ship, which lay to leeward of the cutter so the smacks could easily run down to her after collecting their boats. The copers – from the Dutch *koper* or *kooper* meaning to barter – were mainly old fishing smacks sailing from continental ports, particularly Holland. One coper seized off Kent, but released because she was outside the three-mile limit, had on board 20,000 cigars, 1,000 pounds of tobacco, 200 gallons of brandy, 100 gallons of gin and 100 gallons of *eau-de-cologne*.

The cheap and fiery liquor supplied by copers was known among

121

During summer calms the lure of the coper was hard to resist. Sometimes fishermen were not only poisoned by the raw spirits it supplied, but also became demented with drink and committed acts of folly and violence.

fishermen as chained lightning. It wrought a sad havoc. When visiting 'the devil's floating parlour' for cheap tobacco – a smacksman was seldom without a pipe in his mouth – it was hard to resist an invitation to join in just one little drink. In summer calms time hung heavily on the fishermen's hands. Fresh water became luke warm and was the colour of pea soup. With no books, meetings, amusements, home life or counter attractions of any kind, fishermen flocked to the coper because it offered the only source of relief from the dreary routine of fishing. Rum and gin were only 1s. 6d. a pint, brandy 2s., and a smacksman could become madly drunk for little cost. The liquor was of the cheapest and worst kind, so potent that it sapped a man's strength to the point that he was incapable of holding a rope. His mind became disoriented and he was half drunk and half poisoned.

The harm caused in the fleets by copers was incalculable. Few men got off as lightly as the skipper who told this story in 1896: 'You might say we was like a lot of barbarians in the old days – boozing like brutes and the hands always getting drownded and nobody caring. Look at them copers, or floating grog-shops, with their poison spirits that burnt yer inside like vitriol. I've set sail with a new pair of sea-boots as cost six and thirty shillings and then gone and exchanged 'em on a coper for ten little stone jars of rum – leastways, rum they called it, but it set up sech a craving when it was all gone that I took and drunk off half a bottle of *eau-de-cologne*. And me mincing about in spring-sides like a woman 'cos I'd chucked away my boots like a fool!'

On board the smack *Royal Tar* several skippers met and got drunk and one drank so much that he lay down to sleep and the others could not rouse him. Half in jest they got the turps can and sprinkled some all over him and set him alight. Their intention was to put the flames out immediately but their efforts were not particularly co-ordinated and the man was burned to death. The master of a cargo ship was flagged down by two crew members of a drifting smack and asked to go on board. 'I found the skipper and mate lying on deck helplessly drunk. The third hand was dead and the fourth had been overboard.'

Another smacksman told a harrowing experience of the consequences of a card game aboard a coper. 'One skipper who lost all his money got two of his men to pull him off to our vessel after our crew had gone to sleep. He was like a raving maniac, standing in the bow of the boat without even a shirt on and waving a huge chopper over his head while shouting horrible oaths and threatening to kill the lot of us for taking all his money. I was the only sober man on board and when I saw him getting close I ran below to rouse the crew but they were all fast asleep in a drunken state. Afraid to stop too long lest the boat got alongside us, for it was a very calm night, I returned to the deck and for about half an hour managed with greatest difficulty to keep the boat's crew from getting on board. Then a little breeze sprang up and our vessel forged ahead of them. If God had not sent that little breeze I feel sure there would have been an awful tragedy aboard our smack that night.'

A smacksman called J. H. Sykes was a boy cook aboard a Hull smack fishing near Flamborough Head on a calm day in January when the coper was sighted, and the boat was launched so the skipper and mate could buy tobacco. 'They were on board an hour and the third hand had to bring the boat back as the skipper was too far gone. They brought with them three or four quarts of aniseed brandy which was carried below. In an hour or so the skipper recovered and there were drams all round. I was a teetotaller and besides I was a boy, so I didn't have any. One bottle was soon emptied and another broached. All hands now had what was called a "jolly" and I was sent on deck to keep a look-out. Darkness came and I went below to get somebody to come on deck, but a quarrel was now increasing to prospects of a fight. The language was horrible so I beat a hasty retreat and continued on deck for a couple of hours. At midnight I tried to get the watch awake but he was too far gone. The scene was indescribable, with broken bottles and cut faces. I went on deck and found the wind was freshening. I let go the topsail haliards and let the topsail run down. The breeze freshened more and the canvas wanted reefing but I was alone so I called the skipper but received only a grunt. I was in a terrible fright as I could see Flamborough Head and the wind was easterly (blowing the smack towards the land). Thus the long dreary winter's night passed. It was bitterly cold and snowing heavily. About five in the morning the skipper came on deck saying he felt half dead. He helped me to tack the ship then went below again and roused all hands. As soon as their eyes were open the bottle was again passed round and before daylight all four were drunk once more. To make a long story short, I was alone on deck about twenty-eight hours before one hand came on deck to relieve me. You can imagine the danger we were in, as steam-boats are very numerous there, and I was inexperienced.'

When the North Sea Fishing Act of 1893 made it illegal to supply spirits to the fishing fleets, and illegal for fishermen to buy spirits at sea, many copers turned to Irish waters and some continued to prowl around the North Sea as late as 1906. When the infamous *Kenan* was captured after an exciting chase by the revenue cutter *Beaver* she was towed into Hartlepool and her crew taken to the police station and charged. At Lowestoft skipper George Veal and his mate Isaac Smith of the trawler *Welfare* were fined £4 and £2 respectively for giving away three-quarters of a trunk of fish in exchange for a quart of brandy to a passing merchant ship. The fishermen had not intended to buy spirits, but they had been asked for fish by the ship which was running short of supplies, and the brandy was given as an unasked gift. However the mate made the mistake of falling down the hold as the smack entered harbour, drawing attention to their drunken state, otherwise they would never have been taken to court.

---

20. *Boarding fish on the Dogger Bank, a terrifyingly dangerous job when not even the stoutest small boat and the most skilful seamanship was a match for the mischief of a curling wave.*
21. *When a boat did founder while boarding fish death was almost certain: few fishermen could swim, none took elementary precautions, and their heavy boots and clothes carried them straight to the bottom.*

22. *Trawling fleet on the North Sea on a calm day, when the beckoning flag of the coper provided the only antidote to long hours of tedium and skippers rowed across to each other for "muttoning" parties.*

The Bethel Ships.

In the summer of 1881 a young missionary and social worker who had been working in London docks took a trip to the Short Blue Fleet aboard a fish cutter. 'Our arrival was the signal for a wild scramble for the empty fish boxes we had taken out,' Ebenezer Mather wrote later. 'Boats manned by fellows as rough, unkempt and boisterous in manners as in appearances put off from all the smacks and our deck soon swarmed with four hundred of the wildest men I had ever encountered. Among the 1,500 hands in the fleet there were perhaps fifty or sixty professing christians, but the great majority were utterly careless and Godless and on that afternoon appeared to indulge in language more coarse, profane and disgusting than usual.'

Royal National Mission to Deep Sea Fishermen collecting boxes.

Capstan Collecting Box for
Home Collections.

Mahogany Collecting Box for
Tradesmen's Shops.

At first hand the young secretary of the Thames Church Mission Society saw the evil work of the coper, witnessed the sorry plight of the many smacksmen in need of simple medical condition, and was moved to his heart by the cheerlessness of life in the fleets. Comparing the fleet to a village on land he wrote, 'The inland village lies snugly at the foot of a range of hills with a river flowing placidly by. The North Sea village is constantly tossed to and fro upon the grey wilderness of the ocean, swept by winds as pitiless as the hand of death. The stationary village boasts for its fifteen hundred inhabitants two churches, two chapels, four doctors, a dispensary, a town hall, a mechanics' institute and a lending library. The cruising village possesses absolutely none of these various advantages.'

Mather raised a thousand pounds from a gentleman supporter and purchased and equipped a second-hand smack which was intended to pay her way by trawling, like the rest, but would be manned by christian crew and in every other sense be an evangelist to the deep-sea fishermen. His aim was a vessel with every fleet which was at once a floating church, a dispensary, library and temperance hall – 'a veritable anti-coper.'

Equipped with hymn books and a medicine chest, and sent on her way with scornful jeers and criticisms from fishermen on the dock-side who were suspicious of the churchly do-gooders, the first 'bethel ship' *Ensign* sailed from Yarmouth on July 7, 1882. One smack-owner had told Mather that his men were sent to sea to catch fish, not to shout hallelujah. A deckhand told him that it was impossible for a man to be both a christian and a fisherman in the fleets. Next morning Mather's little ship joined Messrs Leleu and Morgan's fleet – 'but we had a rather strange reception; many smacks bore up round us, and spoke us, and asked what we meant by that big flag that floated in the breeze at our topmast head . . .'

The purpose of the missionary vessel was that it should be not merely a rendezvous for christian fishermen, but a boon for every man in the fleet. The factor that ensured its success from the beginning was that the crew were themselves fishermen. The answers that smacksmen received when they greeted the new phenomenon with a wary 'What cheer, ho?' came in the language of the Dogger Bank. Soon the first visitors were sculling over to the *Ensign* and the humanising of life on the North Sea had begun.

Knowing that the success of the copers turned on their cheap tobacco, Mather tried to get the British Customs to agree that deep-sea fishermen should have the same privileges as foreign-going merchant sailors and get their tobacco duty free. The authorities refused to contemplate such a proposal so the *Ensign* crossed the North Sea to Ostende to stock up with duty-free tobacco which was sold to fishermen for one shilling a pound, under-cutting the copers by a third. As

Mission smack sailing for the fleets.

men came aboard for their tobacco they were given a warm welcome and a cup of tea. Tracts and reading material were handed out. Warm clothing was provided to men in need. The sick and injured were given rudimentary first aid. Sing-songs and prayer meetings were held, on deck or in the converted fish hold.

The movement gathered speed with amazing rapidity. Within two years it had split from the Thames Church Mission and was independent as the Mission to Deep-Sea Fishermen. By 1890 a fleet of eleven sturdy little assistance vessels sailed with the fleets, the words 'Preach the word' emblazoned on one bow and 'Heal the sick' on the other. These were eighty-foot vessels with four berths for missionaries in the stern cabin, a skipper's cabin, and a crew's cabin divided from the hold by a removable partition so the whole space could be converted into a church hall. There was storage for a lot of tobacco, books which were thrown to passing smacks in specially made canvas satchels, and

23. Fine picture of a smack towing its trawl in the North Sea.

medicines. At the end of the decade two mission hospital smacks were launched, *Queen Victoria* and *Albert*, each with a sick-bay with ten berths including two swing cots that kept the patient level as the smack heeled and were invaluable for fracture cases. There was a small organ, to help the singing along, and a new kind of rescue stretcher called an ambulance. Strapped in this, a man was handled from smack to small boat to hospital bed more comfortably than by a dozen strong arms. By 1888 four mission vessels carried surgeons but the only nurses were the crew who made up in cheerfulness and gentleness what they lacked in delicacy and sweetness. On her first five-week voyage the *Albert* treated 396 out-patients and 10 in-patients, putting her own crew members on board other smacks to replace the latter.

During fish-boarding operations the mission smack lay hove-to well to leeward of the steam cutter so that smacks wanting to visit could run down after picking up their boats. This was the time when men usually visited the coper. But with their twenty-foot flags, dark blue with the letters M.D.S.F. in white, the mission vessels declared war on the copers, tracking them back and forth through the fleets so they could not do any business, but when selling tobacco lying well away so fishermen could not visit both vessels.

The *Pall Mall Gazette* described a mission smack arriving in a fleet: 'Right through the floating town the *Edward Birkbeck* makes her way to show to one and all that she is among them, and when the end is reached the skipper shouts across to the mate at the helm, "Shove her right round and go back!" Again the sails flap all round like a flight of birds of prey, and again we make our way back to where the vessels become less frequent, in the suburbs of the town, as it were. And there we lie-to and wait. The fleet knows now that the church is open, the flag is flying lustily on the top of the mainmast, the hold is ready for service, and soon the small scarlet boats, after they have delivered their last trunk of fish to the steam carrier, make their way towards the mission vessel, be it only for a chat or a pound of tobacco.'

When the *Euston* joined a fleet in 1886 the coper bore away and left within an hour, and Skipper Snell heard one of its crew shout to a smack, 'You're all right now, the mission ship's here.' Snell later reported in a letter posted by the steam cutter: 'You will be delighted to hear that we had hardly hove-to in the fleet before we had some forty men on board and we had a grand time. I was forced by the men to stop longer, so we stayed another day.' Skipper Goodchild wrote from the *Thomas Gray*: 'There are no less than eleven or twelve fleets about here, all three or four miles apart from one another, and it is warm work to keep them supplied. The devil is very busy here and in earnest, for the weather is so fine and all the crews can get together.' The presence of the 'navy of the Lord' in the fleets was a steadying influence on the men. When the *Edward Auriol* joined the Red Cross Fleet her

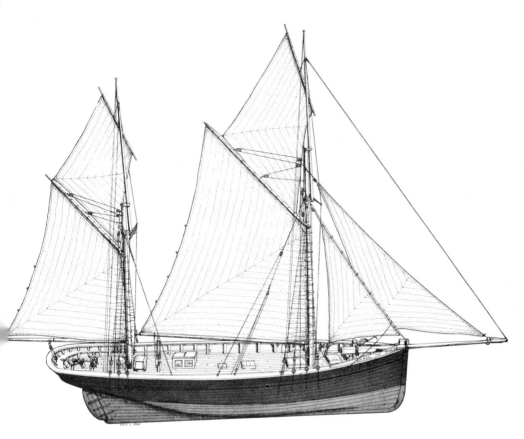

Albert
120-ton mission smack, one of the finest of its type, 1889

skipper ran up the mission flag at daybreak – 'and behold, there was
the coper close to us. A sharp eye was kept on him, and I am happy to
tell you that only three boats boarded him whilst our vessel was with
the fleet. I gave away about three hundred tracts and books and ten
men came for medicine. Poor old coper! I can see him now, running
away to windward as if to the Great Northern Fleet, but the *Salem* is
there.' Also, the mission smacks were sorely missed when a fleet was
left unattended. The manager of the Great Grimsby Ice Company's
fleet wrote to the mission when his smacks were temporarily deserted
by bethel ships: 'There is a coper in the fleet now, and he is doing a
great deal of harm. A lot of men went aboard for tobacco and got that
infernal drink, and many did not get back on board their own vessels,
and some came in, leaving their vessels under-manned. This state of
affairs is terrible. Please to order a mission vessel to the fleet at once,
and let us know when you have done so.'

The simple but hearty evangelism of the North Sea missionaries and their skippers and crews had an extra-ordinary effect on the smacks-men. They were uncomplicated men who led hard and bleak lives. After decades of being out-and-out on the devil's side, they came over to be what missionaries described as out-and-out servants of the Lord. The gospel came to them like water to a man in a burning desert who doesn't know how thirsty he is until he begins to drink. It was the old-fashioned gospel of salvation through the death and resurrection of the Saviour – 'Don't volunteer to be a missionary in the fleets unless you believe in it,' the mission counselled its helpers. 'Modern thought, universal restitution, innate goodness, and so on, won't do in the North Sea. Fishermen want something better than a doctrine of moral patching up: they need a living, loving, personal Saviour, and unless you go out to preach Christ and Him crucified and risen again, you had better stay at home. Fishermen realise only too well what is meant by the horrible pit and the miry clay to be satisfied with any kid-glove preaching which is afraid to denounce sin and point to Calvary as the sinner's hope.' Prayer meetings often lasted four or five hours (with breaks for tea) and thirty or forty small boats would be towing from the mission smack's counter on long painters as she jilled over the waves with helm a-lee and foresail a-back.

In the matter of temperance the mission nailed its colours firmly to the mast. It was allied to the Church of England Temperance Society and every missionary was a temperance worker. Drink was the curse of the North Sea fleets and every Christian knew it. When a man who

A romantic but perfectly true image of the religious fervour that resulted from mission work among the fleets.

was a known drinker went aboard a mission ship he had very little chance of escape until he had signed the pledge. The habits and failings of every man in the fleet were well known, so that once a man had signed there was a good chance he would keep to it. Drinking on the sly was not easy, for a coper did not have a back door. When a man was seen to relapse he was soon visited by his friends. So strong was the temperance movement at Yarmouth and Gorleston that between 1884 and 1891 thirty-one public houses and two breweries closed down, and there was widespread dismay among other publicans because their takings hardly covered outgoings such as heat, light and rent. The Yarmouth Magistrates' Court which had been kept busy for at least two days a week by the drunken behaviour of fishermen was by February 1893 sitting for only about one hour a week.

The fervour with which the mission was welcomed afloat was no less than the zeal with which it was supported ashore. Within two years of Mather's first visit to the fleets the mission had become one of the most fashionable charities of the era in the sense that the plight of the fishermen gripped the imagination of the middle classes. Queen Victoria took an interest, and sent a donation of fifty pounds. Collecting boxes in the shape of capstans were widely distributed and ladies dropped in a donation each time they bought a piece of fish. Ill people who could digest only fish sent money in gratitude. Drawing-room parties in winter and lawn parties in summer, to raise funds for the mission, became annual social fixtures on every middle class calendar. Thousands of people started knitting woollen cuffs, mittens and mufflers that were distributed to fishermen from the mission vessels by the ton. A mission magazine, called *Toilers of the Deep*, was edited by the editor of *Boys' Own Paper* and in 1891 its monthly circulation was over twenty thousand. In 1885 the magazine reported: 'Woollen gifts for smacksmen have continued to reach us during the past month from all parts of the country from kindly helpers in practically every station in life. A domestic servant in Godalming sends us three pairs of mittens with the indication that there are more to follow; a friend at Bridlington and "nine willing little helpers" send fifteen pairs of mittens; a clergyman writing from Blackwall forwards a most useful parcel of mittens, cuffs and mufflers, stating that the mufflers were made by a lady over eighty years of age and that the mittens were knitted by servants of the household. One dear little fellow at Southsea, aged seven and a half, who had received a money gift from his Papa to spend as he liked, promptly sent sixpence to the mission. From Notting Hill comes a most acceptable and encouraging parcel containing among other things nineteen mufflers knitted by messengers at the local post office while awaiting their calls . . .'

Not all knitting was quite right: one jersey which reached a smacksman was large enough for five members of his crew to wear at the same

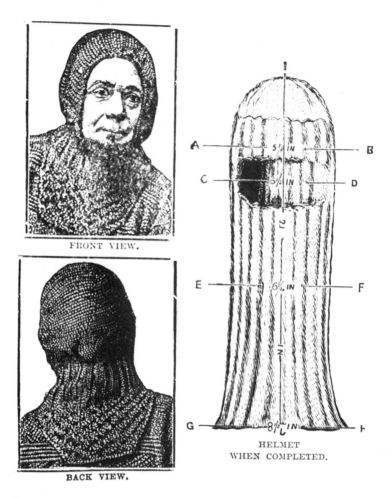

FRONT VIEW.

BACK VIEW.

HELMET
WHEN COMPLETED.

Thousands of woollen helmets, cuffs, mittens and other garments were knitted by volunteer helpers and distributed among fishermen in the fleets by mission ships; patterns were printed monthly in the mission magazine, Toilers of the Deep.

time. Every issue of *Toilers of the Deep* printed instructions for making steering mittens ('the bitter cold experienced in grasping a tiller may be more easily imagined than described'), wrist cuffs which prevented sores and sea-boils, balaclava helmets, seaboot stockings, and mufflers ('to keep out the wet when packed inside a fisherman's oily'). There were also instructions for sewing canvas library bags in which books were thrown aboard passing smacks, and finger-stalls used to protect injured or poisoned fingers which took a long time to respond to treatment because the skin was so thick and leathery, and constantly wet. The commonest seat of poisoned hands was where the finger joins the palm and this could not be protected by a finger-stall so a special mackintosh glove was designed. The patterns were also printed in twopenny booklets which sold by the thousand, complete with prayer for use at work parties: 'We praise Thee for what Thou hast wrought, for numbers of precious souls won to Christ, for many rescued from the evils of strong drink, for all that has been done in relieving suffering and in distributing sound literature on the North Sea . . .'

Less than one tenth of the mission's costs of £25,000 a year were recovered by selling fish. The large sums to buy and equip vessels came from benefactors such as the Duchess of Grafton who gave £2,000, the entire cost of the *Euston* which was named in memory of the late Duke. In late summer every year one mission vessel sailed on a tour of different ports. In each port large awnings covered her deck and she was decorated with bunting. Bazaar stalls were erected on her decks. A brass band played on the quayside and after a formal opening by a bishop or local dignitary the floating bazaar continued for two days with fruit, flowers, books, clothes, and fancy goods made by fish-wives and mission helpers being sold; they did not make a lot of money but were essential as a form of public relations exercise.

For the hundreds of boys at sea the mission launched a letter-writing service, with volunteers inland writing and sending parcels of books, sweets, oranges, lavender bags and other comforts to lads at sea. Extracts of the fishermen's replies were published in *Toilers of the Deep* under the heading Letters From Sea: 'The lavender do smell

Peaceful moment aboard a mission smack in the North Sea.

nice here. I put it in some paper to keep the bag clean till I go home again . . .' 'This reminds us how very acceptable sweet-smelling things are at sea, where the vessel is redolent of fish . . .' 'We have a pet dog and very nice little dog it is. It is a good weather glass for in fine weather it will lie all over the deck, and in a breeze it will not come on deck at all but is always turned in the bunk . . .'

As an object lesson to the fish-eating public the mission bought a condemned smack which had been thirty years on the North Sea and displayed her at the 1891 Naval Exhibition in Chelsea, London. Fully rigged and set in a dry-dock next to a model of the Eddystone lighthouse and another of an iceberg, the *Heroine* was still battered and scarred after her last storm on the Dutch coast which had smashed in her bulwarks and stove in her boat.

Now the mission extended its influence from the North Sea to other waters. A dispensary vessel worked in the Irish Sea, based at Milford Haven, and others worked in the Channel and Scottish waters. In 1892 Dr Wilfred Grenfell, who was the mission's first superintendent, embarked on the most ambitious enterprise yet when he sailed in the *Albert* to what he described as the land of cods, fogs and dogs – Labrador. 'We were glad of the first, sorry about the second and divided about the third,' the young surgeon wrote in *Toilers of the Deep*. Although the Newfoundland cod fishery no longer involved twice-yearly Atlantic crossings it was still a flourishing industry, now supplying mainly the American market to which it was connected by fast steamers and railways. Every June about twenty thousand men, women and children left Newfoundland to sail north in badly equipped and unsuitable boats to the barren coast of Labrador where there was no government, no protection or security of any kind, no spiritual help, and no medical facilities.

Later to be knighted and showered with honours for his work, Grenfell was the son of a chaplain at the London Hospital. He was a

---

24. Mission work: in calm weather a gramophone is taken aboard the Lowestoft smack Suffolk Rose *from the mission ship* Sir William Archibald. *Note the nearest man's long white socks.*

25. Mission work: the "ambulance" by which injured fishermen were transferred by small boat from their smacks to the hospital ship. Note the nails in the soles of the patient's boots.

26. Mission work: delivering mail to a smack's crew temporarily based in Padstow, Cornwall.

27. Mission work: mission smack hospitals treated many injured fishermen, who were replaced on board by the mission smack's own crew.

28. Mission work: doctor treating an injured fisherman aboard a hospital smack; poisoned fingers were by far the most common injury.

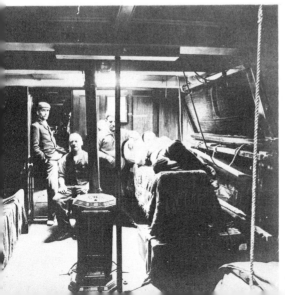

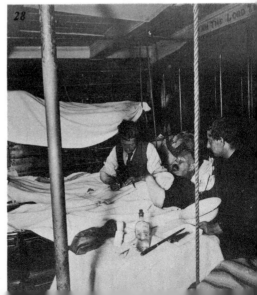

robust character who spent his holidays as a student taking dozens of under-privileged London boys on sailing trips in the Irish Sea aboard fishing smacks which he hired for the purpose. When he qualified as a surgeon in 1887 he was introduced to the mission by Sir Frederick Treves, the senior surgeon at the hospital who had supported Mather and himself had visited the fleets. By the turn of the century he had established several hospitals and nursing stations in Labrador and had the use of a 116-ton steamer, the *Strathcona*, which had been built for his work and equipped with such things as the first floating x-ray machine by an anonymous donor. Just before the First World War the Labrador Medical Mission ceased to be a branch of the Royal National Mission to Deep-Sea Fishermen and was controlled by the newly formed International Grenfell Association. But for five years, from the age of twenty-two, Grenfell was the mission's surgeon and superintendent, and nearly all of that time was spent at sea, on 'active service' – 'These deep-sea fisheries were a revelation to me, and every hour of the long trips I enjoyed. It was amazing to me to find over twenty thousand men and boys afloat – the merriest cheerfullest lot whom I had ever met. They were hail-fellow-well-met with everyone, and never thought of deprivation or danger. Clothing, food, customs, were all subordinated to utility. They were the nearest possible thing to a community of big boys, only needing a leader. In efficiency and for their daring resourcefulness they were absolutely in a class by themselves, embodying all the traits of character which make men love to read the stories of the buccaneers and other seamen of the sixteenth century. There is so much that is manly about the lives of those who follow the sea, so much less artificiality than in many other callings, and with our fishermen so few of what we call loosely chances in life, that to sympathise with them was easy – and sympathy is a long step toward love.'

The copers were not run completely out of business by the mission ships but in the North Sea they were seen now only as shadows. When a coper skipper was persuaded to attend a service aboard the hospital smack *Queen Victoria* he took a temperance card as a souvenir, and one indomitable missionary named Miss Wilkes had herself rowed across to a coper and astonished the fleet by holding a service on board. The mission smacks carried up to four missionaries, some of whom served for short tours, mainly in summer when the weather was not too uncomfortable, and some stayed longer. They were landsmen, ignorant of the ways of the sea, and some spent more time hanging their heads over a bucket than over a Bible. The bulk of the work fell on the mission skippers who were masters of all trades – navigators, preachers, librarians, shop-keepers, medics and, of course, fishermen. However, not all mission vessels carried doctors and the skippers took lessons in simple first aid from a doctor in Kensington.

In 1892 *Toilers of the Deep* published an article giving advice to missionaries planning to work on the North Sea. It provides such a colourful picture of life in the fleets that I have condensed a long extract: 'If, in the fleet, you would win souls, you must abandon all notions of propriety as we know it at home. Suppose you go visiting, that is, taking the Gospel to those smacks whose skippers will not come aboard the bethel ship: if you allow they will give you a civil welcome, will very probably offer you a drink o' tea out of the family mug which has already circulated among the crew, and may possibly invite you to a meal which may be spread on the floor of the cabin. Should cutlery be scarce, as it often is, you will be offered the use of a knife and fork jointly with the skipper. Now you have but one of two choices – either accept the invitation, which is offered in a spirit of true hospitality, and open the way for a friendly straight talk, or decline, and make the way more difficult for a less punctilious man coming after you.

'While on this topic you may very likely have to make the passage either out or home, or both, in a steam carrier. This is the very greatest ordeal. We have crossed the Channel and the North Sea scores of times and enjoyed the trip, but in a fish-screw we always expect *mal-de-mer*. Being of a flat-bottom build, the vessels roll mercilessly on the slightest provocation. Then, if you go to the bows you are met with a current of close air from the foc'sle hatchway. You turn aft and encounter the trawl net which, not having been used for a day or two, has become in the sun anything but sweet. Further aft still is the combined aroma of the engines and of the cook's galley which must be experienced to be understood, and lived in to be appreciated. Everything, as a rule, is very dirty, but if you want to wash take a bucket from the cook's galley, go for'ard and pump half a pail of water, fetch the soap (which, if you are a wise man, will be found in your portmanteau), also your own towel, and there you complete. If you have left the towel and soap ashore you must – well, go without. One of the best questions the mission could ask to ascertain whether a volunteer is in real earnest would be: Do you know what life in a fish carrier is like? If so, are you prepared to voyage out and home in one?

'Don't volunteer unless you have at least broad sympathies with total abstinence. Drink is the fisherman's worst enemy and he knows it. As a consequence, every fisherman in the fleet is a total abstainer. Don't volunteer unless you are determined to fight with the sword of the Spirit, which is the Word of God. We once met two of God's servants out at sea who had gone to do mission work and did not have a Bible between them. They never knew to what extent the omission was noticed among the men; a seaman knows that every skipper should carry his own chart, and not be obliged to borrow one every time he wants to prick off his course.'

141

JOURNAL OF MISSION SMACK *CHOLMONDELEY*, 1888:

October 1: Transferred tobacco to the *Edward Birkbeck* and visited *Clulow;* sailed for Short Blue Fleet and gave away 200 magazines, 100 tracts and sixty pairs of cuffs.

October 3: Seven boats came for tobacco; a coper came to the fleet yesterday but he left again this morning; lent seven library bags.

October 6: Hoisted flag for service, when fifteen met for prayer and praise. It did our hearts good to meet together after the stormy weather, and the power of the Holy Spirit was manifested in our midst; eight boats came for tobacco and six for medicine; one man with smashed fingers and one poor lad came with a burnt foot; visited two steamers and six smacks.

October 7: Visited one steamer and three smacks; two boats came for tobacco.

October 8: Visited three steamers and mission ship *Ashton.*

October 9: The weather today (Sunday) has been very unsettled so we have been unable to hold meetings; although deprived of that blessing, praise God, we have been able to meet in our own little cabin.

October 15: The weather has been very unsettled; Friday and Saturday it blew hard. We shipped a sea on Friday and burst our foresail to shatters – washed away six bulwarks and stove a hole in the boat; but, praise God, our crew are all safe. The Admiral lent me one of his jibs this morning.

October 24: Hauled at 6 a.m; fresh breeze; spoke mission smack *Ashton* and asked them to ferry our fish to the steamer as we were about to start for home; we were told that two poor fellows had just lost their lives, a sea had struck their boat and swamped her. We bore away for home, but finding the wind still freshening we lay-to and reefed; we were among the Leleu & Morgan's fleet all day. It looked like a stormy night and at 4 p.m. I called all hands out to reef. My mate and second mate were standing close to me, and when the work was done I said, "Come along, we will go and get some dinner." I went down into the cabin, and while pulling off my sou'wester the mate shouted out,

---

29. *The broad blue-and-white flag of the Royal National Mission to Deep Sea Fishermen was a welcome and familiar sight among the North Sea trawling fleets in the latter part of last century.*
30. *Smack crews pulling to a mission smack for a cup of tea, a yarn, and a rousing chapel service that sometimes continued for three or four hours.*

"Look out – water!" and the men came tumbling down the cabin one on top of the other. Our little craft trembled and there was a roar like thunder. The sea struck us. I felt our little vessel lay over. As the water came rushing down our cabin I rushed on deck and called out, "Are you all right?" at the same time looking round, when I heard one of the crew say, "Poor Alf is gone." And there, about twenty yards astern of us, was our dear shipmate, floating on the water. I believe that he had been struck with something and stunned, as he did not move when I saw him. I could see him for about two seconds, then he disappeared. The sea had broken our main sheet and main boom, and torn our sail all to pieces. My crew, like myself, stood with tears in their eyes. It was coming on dark so I lifted my heart to the Lord for help and guidance. We cut away the wreckage and pumped out the water which had come through the cabin door. We then went below and met in our cabin to praise God for sparing so many of us.'

The Royal National Mission to Deep Sea Fishermen collection box was familiar in all fish shops.

Over for the Lord.

The Great Yarmouth that David Copperfield thought was rather too mixed up with the tide for its own good was the home port of at least four trawling fleets, but remarkably very little trawled fish was landed there because the catches of the fleets were sent direct to Billingsgate by steam cutter. Also, the single-boaters (trawlers operating independently of the fleets) tended to land their fish at Lowestoft which had better market facilities and was better served by railways. The Rows at Yarmouth were now surrounded by a tide of Marine Terraces, Sea Villas, Prospect Lodges, and Belle Vues which accumulated along the six-mile sea-front, and as far as fishermen were concerned the holiday-makers were welcome to it because they saw quite enough of the sea without wishing to glimpse it from the windows of their cottages.

The sea-front had a fine pier (twice destroyed by ships and rebuilt) and in 1868 it had fifty-eight licensed bathing machines, but male swimmers were prohibited from approaching within three hundred yards of any bathing machine used by ladies. Four years later holiday entertainment commenced on the pier with the Natator – 'the original man-frog' – who performed underwater feats in a crystal aquarium. By the turn of the century there were 100,000 strangers in Yarmouth in mid-season. The fishing for which the town was best known, as it had been for more than a thousand years, commenced at the end of the holiday season when the boarding houses vacated by holidaymakers began to fill with continental fish buyers, curers' agents, and hundreds of lassies from the fishing villages of Scotland (see page 185) whose business was just one thing – catching and curing herrings.

In 1896 there were about a thousand fishing vessels registered at Great Yarmouth and at least half were herring drifters. From the end of September until just before Christmas these were joined by about a thousand drifters from Scotland. Lowestoft had 240 trawlers and the same number of drifters, which were also joined by several hundred Scottish boats. Every night well over a thousand drifters put to sea and set something like two thousand miles of nets which hung just below the surface, like curtains, and drifted with the tide off the East Anglia coast. As the nets were paid out over the starboard side it was the custom for driftermen to doff their caps and say, 'Over for the Lord.'

Aboard the East Anglia sailing drifters, known as luggers whatever their rig, life at sea was no easier than in the trawling fleets but it was very different. The pattern of vessel was the same in the sense that the gear was as heavy, the job as hard. Driftermen sailed the same seas in the same weather, and with the exception of the need to board fish, driftermen faced the same maritime hazards and risks. The main difference was that drifters were constantly in and out of port. While the trawlerman's life was a year-round routine of shoot and haul, with few interruptions, the drift-fisherman's aim was to catch a hold full of fish and land it on the same day so he could get back to sea to catch more. His work was seasonal and he followed the fish, working off the north coast of Scotland in the summer and working his way down the east coast, fishing from ports such as Sunderland and Whitby, until the East Anglia season commenced. Some then sailed round to Cornwall for the mackerel, or converted to trawling and worked as single-boaters during the winter and early spring. The drifterman was less tempted by the devilish coper because he was never at sea long enough to suffer the oppression of lack of company and relaxation. His was altogether a livelier and more varied way of life, though financially it was less dependable.

After the Napoleonic Wars the Yarmouth luggers had three masts and were about fifty feet long with a very long bowsprit. Although they had well-rounded lines, inherited from the Dutch herring buss, the luggers were finer and faster. About 1840 the mainmast was discarded and when railways enforced schedules the magnificent sea-keeping qualities of the rounded hulls were sacrificed for leaner, speedier lines. Luggers were known locally as 'salt carts' because – before the coming of the railway – they stayed out for two or three days, rousing their catch in salt, and returning to port when their holds were full or when the fish they had caught could no longer be kept in good condition; it was usually sold to the smokehouses for making into 'reds' (see page 238). The lug rig was one of the most efficient ever devised; luggers could out-sail and out-point most other fore-and-aft rigged vessels.

---

31. *Crew of the Yarmouth herring lugger* Our Boys; *note the forestay tackle for lowering the mast when lying to the nets and the enormous length of the bowsprit, which in harbour was drawn inboard to prevent damage.*

32. *Herring nets being removed from a lugger at Lowestoft for storage, repairs, or cutching; nets that were left in vessels for more than two or three days tended to catch fire through spontaneous combustion.*

LOWESTOFT.

Its disadvantage was that every time the vessel went about the heavy lugs, or spars, from which the sails were hung, had to be dropped and moved manually to the other side of the masts. In the 1870s there was a revolutionary change to dandy (ketch) rig which lost about a knot of speed but made the vessels more suitable for heaving-to and more handy when tacking, which were important advantages if half the year was to be spent trawling. However the herring boats were still known as luggers, as distinct from smacks which were always trawlers. Many of the old luggers were sold to farmers for as little as two pounds for their timbers, or became houseboats on the Broads.

The sailing drifter of the 1880s was larger than some trawlers, a forty-ton vessel with an enormous mizzen sail that had a very long gaff. The mainsail, dressed with a preservative of horse-grease and red or yellow ochre which gave a distinctive shade of tan quite different to the dark brown or black of Scottish drifters, characteristically had a 'bonnet' or panel along the foot which could be unlaced and removed. This eliminated the real danger of a sea breaking into the belly of the loose-footed sail and filling it with water when the vessel was hard-pressed, and had the effect of taking in a reef. Although outwardly similar there were many differences between a sailing trawler and a drifter. While the smack had a fidded topmast with big cross-trees and a heavy boom, the drifter had a pole mast with no boom and few stays, so that when lying to the nets the mainmast could be lowered into a crutch. The drifter had sweeps, or long oars, stored in lumber irons on the port rail to help her in and out of port in calm weather. There were countless small and subtle differences, such as the drifter's lack of ratlines in the shrouds, which enabled fishermen to identify at a glance the home port of any vessel on the horizon. For example, Yarmouth men rigged a short and upright mizzen mast with a flat truck, but 'Lowstermen' preferred lofty mizzens raked well forward and capped with ornamental trucks.

The 'luggers' with fore-and-aft rig were driven incredibly hard. It was said that an East Anglia skipper never dawdled, was never seen to saunter out of port. He had the sails up and drawing before he cleared the pier and he never looked at his lee rail or cast a glance at his weather rigging – he stood to his big topsail with the white water sluicing along the lee rail and a green arch of water climbing up the forefoot. While the true luggers of the Scots and Cornishmen were at a standstill,

33. *A large Lowestoft trawling smack ghosts seawards as her crew hoist their biggest jib and wisps of smoke emerge from the galley funnel where the boy cook is undoubtedly brewing the first pot of tea of the trip.*
34. *A sight that will never be seen again—scores of fishing smacks crowding the docks at Lowestoft and drying their sails on a balmy Sunday.*

labouring in the troughs of the seas while crews struggled to move the lugs and sails round to the leeward sides of the masts, the Lowsterman was through the eye of the wind in a few seconds and with hardened sheets driving away on the opposite tack. It was also said that an East Anglia skipper was notably reluctant to set his jib, for without the jib his vessel had just sufficient weather helm for him to lean his back comfortably against the tiller.

During the herring season women who took in visitors during the summer found a different kind of seasonal work looking after the miles of nets used by drifters. Most vessels had at least one spare 'fleet' and the Scottish driftermen often sent theirs ahead by rail. One drifter kept two women busy. The work was done in large airy attics in flint-cobble and brick net houses scattered on the outskirts of the town. The job of the woman 'beatsters' was to search for 'spunks' or tears, cut out torn meshes, and repair them by knotting new meshes using beating needles. The nets were hung from the rafters and the women worked with the light of a skylight or large window behind them. When they had finished the beatsters dropped the nets through the floor to the 'ransackers' who assembled them, putting on the 'norsels' (lengths of line fixing the net to the cork rope). The whole thing was then immersed in cutch (the bark of the East Indian betel-nut palm, *areca catechu*) in a big tanning copper in the yard, then spread on the denes to dry, or hung from the net-house balconies. After being 'barked' or 'cutched' the white cotton nets were a dark neutral brown and the fibres were less floppy.

The original 'rough nets' made of twine were ninety-three feet long but gathered to form an actual working length of sixty-three feet. Development of machine-woven cotton nets in the 1870s, lighter and more efficient, allowed much longer lengths of net to be used. These 'Scotch nets' were 150 feet long set to a working length of about ninety-five feet, and in the 1890s drifters set more than a hundred of them, forming a continuous wall of net some two miles long; English boats always set an odd number, such as 101, the extra one 'for luck.' The nets were all joined to a thick rope called the warp. In English boats the original 'rough' nets hung below the warp which was supported near the surface by corks and small water-tight tubs like half barrels which were called bowls. This allowed the nets to stream with the tide, like washing blowing in a wind, and when they were full of herring the nets tended to float to the surface and you could see thousands of herrings trapped in it with their noses showing above the water. The Scottish nets, which English boats adopted, were suspended to a lighter warp that hung below the nets; the nets were kept buoyant by dozens of canvas buoys, called pallets, that were painted different colours and numbered. This allowed the net to belly under the weight of fish, like a sail curving in the wind, and if a steamer happened to

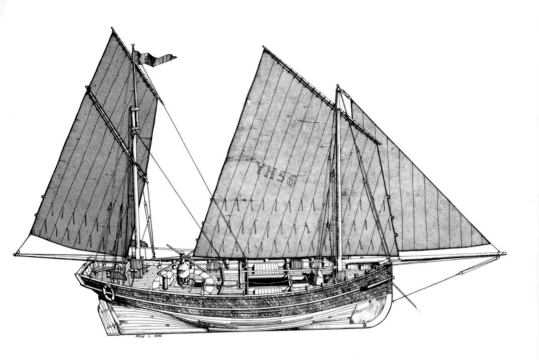

Gipsy Queen
20-ton Yarmouth lugger, 1858

pass through the nets there was less danger of the warp being cut by
its screw. If the nets were not used for two or three days they were
liable to overheat and had to be brought out of the drifter's net-room
and thoroughly wetted down. In July 1888 the herring yawl *Countess of
Fleetwood* was lost nine miles west of Barrow when her nets caught
fire by spontaneous combustion.

Like smacksmen, herring fishermen wore a primitive and picturesque
uniform. Their jerseys were almost invariably white. Over them were
jumpers of stout canvas which were barked like the nets and sails, and
had no buttons to catch in the nets and drag a man overboard. The
skippers, who had more standing around to do than the others, favoured
a long night-dress style that extended down to the ankles and kept
out the draughts. In winter driftermen wore a kind of sheepskin
waistcoat turned woolly side inwards and dressed with linseed oil. Rough
home-knitted woollen stockings were tucked into leather boots, thigh-
length in winter, half-length in summer. There was a marvellous
variety of head-gear, including large sou'westers, varnished bowlers,
top-hats, and cheese-cutter caps.

Until the 1860s luggers were crewed by eight to twelve men of whom
between two and four were known as joskins. Often farm labourers
from inland with no seafaring background, these were hired for the

SOUTH BAY, SCARBOROUGH FROM THE SHORE ROAD 6943

strength of their arms and backs; usually they did not share the cabin with the fishermen but ate separately on deck and snatched a few winks during the night by finding a bed where they could, on nets in the hold or in the lee of the bulwarks. Their job was to assist in the hardest work aboard any herring lugger: hauling the nets. This required a weary tramp round and round the capstan, each of the joskins leaning with all his strength on a seven-foot capstan bar. Depending on the length of nets that had been set, their circular journey on the wet and tossing deck covered a distance of between seven and thirteen miles. Later, the rotary capstan was replaced by a patent winch with cog-wheel and pinion that required the strength of only two men, and in the early 1880s this was replaced by the steam capstan, known as the 'iron man'. Riots broke out in the herring fleets in 1884 when owners took an extra share of the catch to cover the capital cost and running expenses of this labour-saving device.

When shooting the nets, about thirty miles out, every man on board had a particular job. The skipper took the helm and sailed the lugger slowly across the tide. The young cook dived into the warp room to ensure the rope did not kink or foul as it ran out, and he had to look lively to prevent his legs being caught. As the nets were passed up from the net-room by the net-stower, the net-rope-man and the whale-man paid them over the side. The hawseman, who was usually also the mate, tied the seizings which made the train of nets fast to the warp. At the same time the younker, who was a man of all work, passed overboard the pallets or bowls that were tied to the upper end of each seizing. With the weight of the warp hanging below it, the net came upright in the water, suspended from the pallets. By adjusting the length of the seizings the depth of the net below the surface was controlled. After about half an hour the last net was over the side and as the lugger came up into the wind the long wriggly line of floating buoys extended for up to two miles beyond her bows. The mizzen sail was sheeted in hard amidships to keep the bows into the wind and the mainmast was lowered into a crutch, or mitchboard, about eight feet above the deck. On this was fixed an octagonal lantern, with two candles, which looked like an old-style street-light. About one hundred fathoms of extra warp was paid out and a spring-line called a tissot was rigged to take the strain of the drifter's snubbing to the warp as she rose and fell in the seas.

35. *A busy scene at Scarborough, about the turn of the century, as smacks and cobles land their catches on a calm morning; hotel-owners complained that the smell of fish being carried through the streets to the railway station caused offence to visitors.*
36. *Auctioning a small catch on the quay at Whitby, c.1880.*

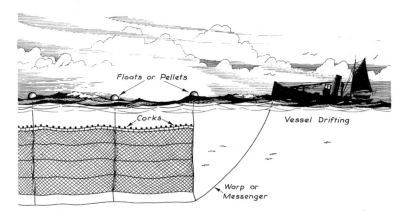

Floats or Pellets

Corks

Vessel Drifting

Warp or Messenger

K.C.Lockwood

A drifter shoots long lengths of nets just below the surface at right angles to the expected path of the shoals so that shoals of herrings are enmeshed by the gills.

Now, if the cook was up to the mark (and look out if he wasn't!) it was dinner time and all hands but one dived below into the tiny aft cabin where the galley stove was roaring on its flagstones. Food was good and more varied than in trawlers because fresh supplies were always available. Large square tin dishes containing a leg of mutton, gravy, potatoes and dumplings were put on the floor among the sea-boots which stopped them sliding around and you helped yourself.

In the hoodway – the curved shelter over the top of the ladder giving access to the cabin – every prudent skipper carried a hatchet with which to cut away the nets in an emergency, and a flare made from a thick bundle of cotton waste on a handle which was stored in a small conical tin half filled with paraffin. If a steamer bore down on the drifter the flare was snatched dripping from its tin, thrust into the galley fire to get it going, then taken on deck and waved as a signal. In fog, a brass horn was fitted to the muzzle of the fire bellows and sounded every minute or so by the man on watch. A speciality of Lowestoft drifters was the shoe kettle – a tin kettle shaped like a very large pointed boot with a handle on the 'heel'. This was filled with water, tea, and sugar, and thrust into the red-hot embers of the

37. *Speckled with silvery fish scales, fishermen unload herrings into quarter-cran baskets at Lowestoft.*

154

capstan-engine boiler: it produced a mug of hot tea for each man on board in less than two minutes.

Around midnight the first one or two nets were hauled in to see if they contained any fish. When the swim came a drifter could catch so many herrings that the nets sank to the bottom and had to be recovered with grapnels called creepers. If the wind completely reversed its direction the drifter might have to let go her nets and sail round to the other end of the fleet, because the vessel could work only if she was down-wind of the nets. Unless the first 'look on' showed a lot of fish in the nets the long haul commenced at about 3 a.m. and the men would be lucky to get all the nets in by the first grey streaks of dawn. As the joskins tramped laboriously round the capstan, hauling in the warp, the seizings and buoys were removed from the nets as they came aboard. Then the nets were stretched flat between four men standing on either side of the fish-room hatch and given a hard shake so the fish fell out. This 'scudding' was very hard work, and the men often changed positions. Down in the warp room the cook had a tough time, as fisherman John Gawne described in 1895: 'At midnight the cook was up again to coil that confounded rope, as thick as my wrist. You pulled it through a hole in the deck and coiled it down, good work if it was done in four hours. I have been down there seven hours getting it in good and tight against the deck. I don't know how fellows who were seasick did it. When shooting you had to watch the rope going out, and I would rather coil than watch; lots had their legs badly hurt. For five nights a week the cook coiled down the rope every night, the only light a little swinging lamp with no glass in it, and the smell of the lamp and the anchor chain! Jelly-fish (barmoos we called them) made the rope slip out of your hands and if you got them near your eyes look out! They were awful, octopuses too, and we would be hanging on, boat rolling, rope slipping!' A really good haul of herrings filled the hold and fish lay knee-deep over the decks in a silvery tide. As the fishermen walked on them they squeaked like kittens, and as the air escaped from their swim bladders they made a noise compared by one fisherman to that of all the mice on the pharoahs' granaries.

Now the lugger turned for home, cramming on all sail because first boat in earned the best prices. As light stole over the dark sea other boats all round her would be doing the same. After his long and painful toil in the warp-room the cook now had to be quick and nimble because once his work was done a fisherman was impatient for his breakfast. The cook scooped up a bucket of herrings and immediately began to prepare them. With six or seven men aboard, each having up to a dozen herrings for his breakfast, he had a lot of frying to do. Walter Wood described the scene: 'The skipper takes the helm till breakfast is ready then, willingly obedient to the summons, we tumble below again and fall hungrily upon tea, bread and butter, and herrings – herrings freshly caught, gutted, beheaded, deprived of tails, slashed with jack-knives latitudinally so that when a huge dishful of them is placed on the floor, piping hot from the boiling fat in which they have been fried, we can bend down and help ourselves, and with our fingers strip the crisp, delicious morsels from the bones and eat them.'

Steam tugs stood by outside the harbour entrances as the drifters hove in sight early in the morning. As they tied up at the quayside their skippers jumped ashore and ran with sample buckets of herrings to the buyers. The first ones in soon found themselves hemmed in by scores of others until drifters lay ten or twelve deep along the quayside. In Yarmouth they moored bows-on to the quay to leave as much room as possible and in the strong currents of the River Yare there was a constant grinding of gunwales and upper works as vessels rubbed together. In Lowestoft's square-shaped basin there was no such order. Armed with a speaking trumpet, the harbourmaster shouted instructions, getting vessels away seven or eight at a time because incoming vessels did not hang back to let others get out and create more room, and he carried a chopper to cut away any mooring ropes when necessary. Men were fighting for their living and it was every man for himself – 'Dog eat dog, without question,' as one old drifterman described it.

The Real Price of Fish

A man who made his living on the Dogger Bank carried his life in his hands. On the North Sea life was cheap, and until 1883 a death did not even have to be reported unless it also involved damage to the vessel. Collision was a perpetual danger, particularly when a snowstorm scattered a fleet. Vessels had low bulwarks and if a man went overboard it was the end of him. In fishing towns every winter storm left in its wake a harvest of new widows and destitute children. The lives taken by North Sea storms make lengthy casualty lists.

   1887 (March 3): 36 smacks lost, 215 men and boys drowned, leaving 88 widows, 164 children.

   1889 (February 9): 94 fishermen lost (51 from Grimsby).

   1890 (June 25): 42 lives lost.

   1891 (November 11): 42 lives lost.

   1894 (December 22): 146 men and boys lost, including 106 from Hull who left 55 widows and 128 children.

But one of the worst of all was the 'Black Monday' storm that swept down on the North Sea fleets, with no warning but the natural signs in the sky, on March 4, 1883. Its toll was forty-three smacks lost with all hands, seven abandoned, eighty-nine damaged – and 255 men and boys were never seen again.

'Let those who speak of the price of fish spend one winter's night aboard a trawling smack,' wrote Ebenezer Mather, founder of the Royal National Mission to Deep Sea Fishermen. 'You must be prepared for keen frost, blinding snow, and all around the grey wilderness of a floating ocean swept by winds as cold and pitiless as the hand of death. On the North Sea a snowstorm is not merely regarded as objectionable but is dreaded more than any other visitation. When a snowstorm bursts upon a fleet of fifty to two hundred trawling smacks

38. *Pumping for their lives—fishermen fighting a North Sea gale as their sails are torn to ribbons. Many vessels were swept clean by huge waves.*

it strikes dismay in its advance and leaves desolation in its wake, for collisions are inevitable and involve sad loss of life.' In winters that froze the Thames from bank to bank North Sea smacksmen sailed through ice-floes, and had to boil kettles to unfreeze their mainsheet blocks and rudder pintles.

Collisions were frequent in fog and winter squalls. Many fishing boats with poor and flickery navigation lights, if they had lights at all, were simply cut down by large sailing ships and steamers which might not even feel the bump; sometimes there was an opportunity to scramble aboard the other vessel as the two craft lay momentarily locked together. The mission smack *Edward Auriol* found a hulk drifting near the Belgian coast in thick fog. 'I could see she had been run into by a steamer,' the skipper reported, 'for she was cut half-way through. It must have been an awful state of things for the crew – a crash, a scream and a plunge.'

Incredible tales were told of rescue and survival in North Sea storms. Distressed smacks drifted so close to each other that they were in danger of colliding, but neither could help the other. A man was swept off the deck of his own smack by a wave and deposited on the deck of another. In maelstrom seas fishermen saw other smacks drive into a wave and emerge clean-swept, without bulwarks, boat, masts or spars, and usually without the men who had been standing on the deck. One crew found themselves up to their waists in water in the cabin, the fire and lights out and the vessel on her beam ends. When they tried to get out they realised the heavy trawl gear had become jammed across the hatch. They were trapped and the vessel was sinking. The skipper climbed on the table and tried to break through the sky-light. There was no time to grope through the icy water for the hatchet so he attacked the woodwork with a pocket knife and eventually succeeded in making a hole through which they all escaped to find everything on deck had gone – masts, boat, tiller, everything except the trawl.

The worst gale Skipper John Mann experienced in a lifetime fishing under sail came at four o'clock in the afternoon on October 29, 1880 when he was skipper of the *Willie*: 'We lay with our head to the east'ard until midnight, then we tacked. My boy Jack was third hand, and he had first watch. "It blows hard but we have a good ship and she lays well," he said, when I went up to see him. After supper the watch was rigged to go on deck when a heavy sea struck the vessel and hove her down on her beam ends. The next moment I felt myself under water, struggling for breath. It seemed a long time but I suppose

---

39. Storm at sea! Wives and sweethearts wait on the pier-head with bated breath as one after another the small ships of the fishing fleets run to safety.

it was not many moments before the vessel righted. My first thought was for the crew and I called everyone by name and they all answered.

'The light had been put out by the water, and telling everyone not to move I groped in the dark for some matches and damp lamp cotton that I always kept in a bottle on top of the cupboard. Meanwhile, instead of obeying orders and keeping still, some of the crew went on deck. Jack sang out, "Hand up the hatchet, the gear is overboard and if all the trawl runs out it might bring us up stern-first and swamp us." When we found the hatchet one of the crew called out to those on deck but got no answer. I told him to call louder, in case they were for'ard, but there was still no answer.

'I then went on deck with the lantern but could not see anyone. We had lost three hands, and could not make out how they went. I was overcome. In all the gales of wind, and all I had passed through, I never lost my nerve or presence of mind. But then, although I had checked that everybody was safe, we lost three hands. I trembled like a leaf and for half an hour couldn't do anything, even if it had been to save my own life. After a time I thought this wouldn't do, if I gave up the others would. There were four of us left out of seven. I said, "Now my lads, we must pull ourselves together: let's see what we can do to save the ship."

'Both main and mizzen masts had gone, the boat had gone, the companion all gone, nearly all the bulwarks smashed away, and the stanchions along the weather side were gone. The sea had thrown the trawl gear from the weather side clean over the vessel and the warp was running out to leeward. The first thing we did was cut the warp, then cut the shrouds so the mast trailed by the forestay. The main thing was to get the water out of her for she was more than half full. We couldn't find any pump gear but we had a bucket and an old kettle so we set to work to bale out the water. By noon the cabin floor was clear. All day we never saw a vessel but near sunset a Hull smack took us in tow. It was a dreadful four days getting to the Humber for we had nothing dry on board and we were in wet clothes all the time. We could not keep a fire because the funnel (chimney) was torn out of the deck and we had to nail canvas over the hole, but thank God we got to Grimsby all right and soon got dry clothes. Then there was the going home to the wife and daughter. I thought they would go crazy when they heard of the loss of the only son and the only brother.'

If any of the North Sea gales which caused such great loss had

---

40. *Some of the last of the sailing smacks alongside the quay at Ramsgate in 1922; these vessels survived between the wars because they caught fish of very good quality, though not much of it, and landed it fresh and not iced. Most of the fish landed by stream trawlers was fit only for frying.*

occurred ashore in a colliery or a railway accident there would have been national grief and lamentation. But worse than the tragic storms was the steady loss of fishermen by ones and twos. Again, if every accident causing loss of life occurred on the same day there would have been a national outcry: it would have been a disaster of massive proportions. But because fishermen were killed at a steady rate throughout the year their passing attracted little attention. In January 1887 a Grimsby newspaper reported: 'The new year has opened unpropitiously with seven lives lost in the first fourteen days. On the second, Robert Baker, deckhand of the trawl smack *Rowland*, aged eighteen, native of London, was lost at sea; on the third, Frederick Blackbond, an apprentice, was drowned from the *John and Sarah Ann*; on the fifth, A. T. Mitchell, aged twenty-four, deck-hand of the *Martha Somerville* was drowned; now the Grimsby smack *Henry Freeman* has arrived in port and the master and owner reports the loss overboard of the second hand, William Dann, aged twenty. Mr Mundahl's cod smack *Oscar* has also come into Grimsby and the master reports that on Tuesday afternoon a heavy sea swept away two apprentices; attempts were made to rescue them but they sank and drowned . . .'

Once a man was over the side there was little chance of rescue. Sailing vessels were not easily stopped or turned round, and if the man was to windward getting back to him against the wind was a laborious and difficult process. A man in the water floats with barely his head showing, and in seconds he can be lost to view. In darkness he has practically no chance at all, even if he is not wearing heavy clothing and long boots which drag him down straight away.

Not until 1883, when a fishing skipper was obliged by law to report deaths and accidents of all kinds to the Board of Trade, did a clear picture of casualties emerge. In the ten subsequent years 2,389 men and boys were lost from fishing boats registered in England and Wales of whom forty-eight per cent went down with their ships. 600 fell or were washed overboard (about one a week); 33 (mostly boys) were dragged overboard by buckets while drawing water from the sea; another 200 were knocked overboard or killed by the ship's gear; 165 were drowned from small boats, mainly while trunking fish; 90 died of natural causes at sea; 27 were known to have committed suicide.

Between 1876 and 1881 some 425 fishing vessels were lost with 866 lives; between 1884 and 1888 a total of 1,328 men and boys were lost. From 1880 to the First World War 1,066 English and Welsh fishing vessels were lost of which nearly a quarter were from Great Yarmouth; one third of them foundered, one third were lost in collision, and the remainder were stranded or burned out. Blackest year on record was 1894, when 492 fishermen were killed – about one man in every 60.

Storm warnings had been issued by the Royal Navy since 1841 and telegraphed to the main ports, being indicated by signals hoisted on

Sighting a familiar sail homeward bound.

flagpoles at the harbour entrances. But local telegraph stations closed at eight in the evening and did not open before nine in the morning, by which time any smack going to sea had already sailed. Large barometers were erected in many fishing villages but on the whole they were used by fishermen only to confirm their own instincts and at other times were ignored. These were of no help to fishermen at sea, but on the rare occasions that a smack did carry a small barometer the question it was asked was not how much grace was available for the smack to reach safety, but how much time it had to continue fishing.

Regulations were made to compel smacks to carry lifebuoys and lifejackets but no Act of Parliament could make men careful, and it was recklessness and carelessness that led most men to their watery graves. In 1893, a year in which 329 fishermen were lost in England and Wales (nearly one a day), the report of the Inspector of Fisheries stated: 'I regret to have to add that that in eight cases the lifebuoys were below or stowed away, or even in a cupboard, when the accident occurred, and was therefore of no use. There still appears to be a dislike among some fishermen to wearing lifebelts when fish-ferrying in rough weather.'

As a rule fishermen did not make provision for their wives or children left behind. Many fishermen subscribed a few shillings a year to the Shipwrecked Fishermen and Mariners' Royal Benevolent Society which had been formed in 1839 but benefits were limited. For a premium of three shillings a year a man who had been a member for more than five years received £2 for loss of boat or clothes (not nets or gear), his widow or dependent parents £4 if he was killed, a widow with four children £8. In 1884 the Royal Provident Fund for Sea Fishermen was founded by the Prince of Wales and branches were opened at all the main ports. It was modelled on the similar scheme for the coast-guard, in which men paid six to twenty-four shillings a year so that dependents would receive on their death sixteen to sixty pounds. In Scarborough at the turn of the century the fishermen's mutual insurance society was dissolved because they found that impoverished dependents could do better by appealing to the charity of visitors.

As every battle has its heroes, every North Sea storm produced a crop of brave deeds, most of which went unseen and unrecognised by the public. Fishing smacks were the lifeboats of the offshore waters and the lives of hundreds of sailors and passengers, British and foreign, were saved by fishermen who considered it all in a day's work. One smack set out from Yarmouth three times and each time was back within a few days with a crew of a wrecked ship; the skipper was heard to mutter that he had been warned by his owner to stop catching Dutchmen and bring home a few fish. In ports other than the main fishing centres, such as the Thames estuary, fishermen made half their livings from salvage and life-saving. Smacksman Tom Barnard of

166

Rowhedge, Essex, was reckoned to have rescued 900 people between the 1830s and 1881. The Board of Trade paid compensation to fishermen who lost gear, or had to jettison fish, or missed their market, as a consequence of rescuing lives, and many fishermen were awarded medals for gallantry. Men who had plucked foreign seamen from sinking ships were presented with medals, diplomas, gold watches, binoculars and cash by grateful foreign governments.

Fishermen were fatalistic about death. They were quick to dash to the rescue of others, but slow to take preventive measures for their own sakes – 'We're used to being drowned.' Outside the fishing ports hardly anyone buying a piece of fish knew anything of the drama that lay behind the catching of it, or suspected the price that had already been paid for it.

MEN LOST OR KILLED AT SEA FROM FISHING VESSELS REGISTERED IN ENGLAND AND WALES

1884	300	1894	492
1885	218	1895	333
1886	178	1896	152
1887	334	1897	205
1888	298	1898	239
1889	310	1899	197
1890	311	1900	335
1891	268	1901	204
1892	197	1902	218
1893	329		

IV

THE SEA HATH FISH ENOUGH

Here's health to the Pope,
  May he live to repent,
And add just six months
  To the term of his Lent.
And tell all his vassals
  From Rome to the Pole,
There's nothing like pilchards
  For saving the soul.

(Cornish fishermen's toast)

Fifie Skaffie and Zulu.

Compared with English fishermen employed in the great North Sea trawling factory, Scottish fishermen were small country shopkeepers. Each Scottish fishing boat was the focus of a small family business in which wives, children and grandparents played an essential and hard-working role. In good times the tradition of individual ownership brought a hard though fulfilling way of life, never rich and seldom prosperous, but potentially comfortable and always independent which suited the Scotsman's nature. But the tradition also legislated against co-operative effort, and in bad times the Scottish fisherman fared worse than most.

The nineteenth century, however, was in general a 'good' time. Most of Catholic Europe now bought its herrings from Scotland; thousands of barrels were railed deep into Russia and herrings, eaten with potatoes, were the staple food of peasants. Every creek, every little port, had its fleet of small boats which fished for herrings during the season, and for the rest of the year long-lined for white fish that was salted or smoked because, unlike England, there was no organisation for distributing and disposing of it; trawling was not known at Scottish ports until the end of the century. Many boats were owned by crofters, who divided their time between the sea and the land. Others fished full-time, following the herring season to different ports which were served by rail. In September 1858 *The Scotsman* reported from Dunbar, known as the Yarmouth of Scotland because its curing houses specialised in smoking red herrings: 'Forward to nine o'clock every boat that has reached the harbour mouth is laden to the gunwale, and in the distance are seen an immense number of boats making for here and the Fife coast, apparently heavily laden. Prices commenced at 27s. 6d. (a cran) but are now 20s. Such a day's fishing has not been seen at Dunbar since 1823 when the fish were sold at one shilling a cart-load for manure. From the resources of the railway and the "empties" (boxes) in the hands of buyers, the whole will this year be secured for human food.'

The fact that the Scottish herring fishery could reach such a pinnacle

171

of vigour when harbours were inadequate and boats were unseaworthy and unsuitable, is a comment on the fishermen's remarkable reserves of persistence and amazing lack of common sense.

Scottish fishing boats, low in the water, flattish, clinker built with one or two square-shaped lugsails, evolved from Viking longboats. Bows and stern were both pointed, which was a considerable space-saving advantage in harbours like Wick that were packed with hundreds of boats during the summer season. The boats were very fast under sail but unwieldy and difficult to handle, and travelled long distances under oars alone. They were completely open, with no decks, there was no shelter for the crew of three to five, who napped under a tarpaulin.

In 1820 the six hundred boats fishing from Wick in the summer season averaged 148 crans of herrings each. A fisherman used about twenty-five small nets with an area of 4,500 square yards, and he could practically carry them on his back. By 1860, with the advent of machine-made cotton nets, a fisherman set about 16,800 square yards of net from the same boat, and he needed a cart and a strong horse to carry them, but his average catch was down to less than a hundred cran, and in 1863 it was only seventy-nine cran. This decline was due partly to the fact that herring shoals were moving further from the coast, but mainly because the fleet was able to fish for only about twelve nights of the season. The small open boats were so unseaworthy that they could neither follow the shoals into deeper water nor cope with stormy weather.

At the beginning of the century a typical fishing boat was only about twenty-two feet long but it was capable of being beached easily, or finding shelter in tiny rock coves and creeks that abounded along the coast. By the 1850s boats were more commonly forty feet long, displacing about five tons. The bigger boats required larger and deeper harbours, which were scarce: if caught at sea by a sudden storm there was nowhere for them to run to. On the entire coast of Fife, where there were 420 boats and 2,100 fishermen, no harbour could be entered at low tide. The same was true on the south-east coast between Edinburgh and Berwick. At Dunbeath the harbour was, typically, only a creek slightly protected by a wall of loose stones built by fishermen. Peterhead was the only all-weather all-tide port of refuge between the Firth of Forth and the Moray Firth.

At Wick, the only all-weather port on the coast of Caithness, one hundred feet of Telford's fine new outer quay or breastwork was breached by a storm in 1827. Repairs were completed in seven years but the masonry was never removed from the habour entrance because contractors said it would undermine the foundations. The result was that for five hours in twelve the harbour was inaccessible to loaded fishing boats.

Inevitably it took a calamity – 'unprecedented in the annals of

British fisheries' – to bring the situation to public notice. On August 19, 1848, at the height of the herring season, more than six hundred boats put to sea from Wick and three hundred from Peterhead. The weather looked threatening as they shot their nets in the evening. By midnight wind and sea had risen to a howling south-easterly with heavy rain, and many boats ran for shelter. At Wick there was only five feet of water in the entrance and in the heavy seas that were raging no boat could enter. Although the streets of the town were gas-lit the only light for guiding boats into the harbour was a lamp which blew out. A similar disaster had occurred three years before, but there were still no ropes, lifebuoys or ladders available for life-saving. Twenty-five men died in the harbour entrance that night, and twelve were lost when their boats were swamped at sea.

Harbour-master Peter Taylor, fifteen years in the job, put the whole blame on the fishermen who, he said, had ignored the rapidly falling barometer. Emigration agent John Sutherland, who had been recruiting settlers for America, saw boat after boat dashed against the breakwater and himself rescued two men. He told the subsequent inquiry: 'I have spent most of my life in what may be erroneously called the wilds of America, but I beg to observe that civilisation is not so far behind there that no attention is paid to the preservation of human life.'

At Peterhead the harbour entrance was only thirty yards wide and obstructed by badly moored steamers that occupied two thirds of it. There were no lifelines or hauling lines on the quays and valuable time was lost as rescuers leapt aboard boats in the harbour to hunt for ropes. In a few hours 30 fishing boats were lost, 33 were damaged or stranded, 31 men perished. Small ports scattered along the east coast also suffered. At Dunbeath a spate washed away the stone breakwater and many of the village's 106 fishing boats were pounded to pieces by the surge. The total cost of the storm was 124 boats lost, 100 men dead, 47 widows and 161 children destitute.

With commendable and unusual speed Parliament ordered an inquiry. The Admiralty appointed Captain John Washington to the job and he began hearing witnesses at Wick within six weeks of the disaster. He found the 'grand cause' was the lack of good harbours; only £2,700 a year was spent on Scottish harbour works. He also said that the fishermen should have ridden to their nets longer – at least until they knew there was sufficient water in the harbour entrance. At Peterhead many boats had got into trouble when making their final approach to the harbour entrance because their crews, of which an average three in five men were landsmen, were not sufficiently skilled to make a quick and skilful job of dipping the lug; this resulted in lack of control, and the boats sagged away to leeward and missed the entrance.

But the nub of the problem, which the perceptive Captain Washington identified, was why did the boats have to run for shelter in the first place – why were so many unable to ride out a storm that might have been unseasonable but was by no means uncommon? Washington pointed out that Scottish fishermen unlike their counterparts in the South and West of England, insisted that the only boat fit for herring fishing was an open one. Among Mounts Bay men the feeling of anxiety over their own safety in the event of an on-shore gale springing up was simply not known. Yet some Fraserburgh boats already had a forecastle deck which kept the water out of the bows when meeting a heavy sea, so why did not other Scottish fishermen follow suit?

The answers came from the Scottish Board of Fisheries. Its secretary, the Hon B. F. Primrose, claimed that decked boats were too heavy to row against light winds and it was essential to deliver fish to the curers within twelve hours of its being caught; while the boat was rocking with no steadying canvas when the lug was dipped the men could grip the thwarts with their knees and were able to keep their weight low in the boat; stowing nets in a hold would be a disadvantage because they had to be landed and dried daily; if the boats had bowsprits and bumpkins, like the 34-foot Scarborough yawls which Washington recommended, they would not be suitable for crowded harbours. It might have seemed a reasonable argument at the time, but the result was that the dangerous and inadequate open boats persisted in general use for another twenty years.

The first decked boat was built at Eyemouth in 1856 and by 1872 there were forty decked boats in the district, each one about fifty-six feet long. Despite the obvious advantages these were only a small proportion of the total fleet. At Coldingham and Eyemouth, where nearly all boats were decked by 1872, the average gain from herrings was up to £550 a boat; at nearby Burnmouth, where boats were nearly all open, the average gain was only £160.

In 1867 the National Lifeboat Institution became alarmed at the frequent and unnecessary loss of life by Scottish fishermen in open boats, and 'the coast boatmen being an inert class, not readily departing from what they have been accustomed to' its committee designed and built four or five forty-foot prototype decked boats which were lent or hired to experienced fishermen at the main ports. Each boat had a water-tight bow compartment which served as a cabin and another in the stern; there were side decks with scuppers around a large hatch which could be covered with a tarpaulin lashed to high coamings. Mainly as a result of this initiative the movement towards decked boats gathered momentum.

One of the advantages of larger boats was that fishermen could now follow the herrings for most of the year, starting in the Shetlands in May and finishing at East Anglia in December, and line-fishing nearer home in the winter. In the Shetlands most of the 1,613 boats registered

174

at Lerwick in 1874 were open boats less than thirty feet long and powered by six oars. These 'sixerns' or 'sexærings' participated in the summer háf (ocean) fishery when, working in pairs for safety, the frail double-ended boats ventured forty miles out to fish for cod and ling with long-lines. Each boat carried two sacks, one of peat fuel and one of stones for ballast. When sailing homeward the catch was protected by a layer of halibut, which then had no commercial value, and on landing the halibut was thrown down as skids on which the boat was run up the beach. The catch was split, cured, and bulk-salted on the shore and stored for export to Spain as stock-fish. By the 1870s the háf fishery was declining due to trawling, and a fantastic boom in herring fishing took its place. In ten years the number of herring boats increased from about fifty to 932, and the barrels cured from 1,100 to 200,117. The last sixerns were used early this century as tenders for steamers carrying peats and livestock among the islands.

West coast fishermen suffered grievously in 1860 when drifting interests managed to promote an Act prohibiting the catching of herrings on any part of the west coast from Cape Wrath to the border between January 1 and the end of May. The reason given was that the circle nets used in Loch Fyne by boats called 'nabbies' working in pairs destroyed the shoals. With abundant food lying at their doors the population of the Western Isles practically starved; herrings could not even be caught for bait with which to catch cod. At Anstruther, Fife, a fisherman could catch and cure any quantity of herrings and have them approved and branded by a government official. But his counterpart at Ardrishaig, on the shore of Loch Fyne, risked imprisonment and the seizure of his boat if he did the same – the mere presence of scales in the boat was sufficient to ensure a man's conviction. These restrictions were ended in 1868 after the whole industry had been investigated by a Parliamentary Commission of which Professor T. H. Huxley was a member.

As east coast fishermen changed over to fully decked and larger vessels, two distinct types of boat emerged. The differences were not apparent in the rig or in the plan: all were beamy boats with sharply pointed ends, carried two masts, and were rigged as luggers. The differences were mainly in the profiles of the hulls.

On the southern shore of the Moray Firth between Portmahomack (just north of Inverness) and Banff, boats had a rounded, yacht-like stem and a very sharply raked stern. These boats were known as 'skaffs' or 'skaffies' and because of their relatively short keels were nimble in confined waters and answered readily to the helm. The rounded bow reduced damage if the boat was inadvertently run over her own nets, but it did not provide much of a grip on the water and the bows were liable to fall off the wind, which in a head sea could be dangerous. The large overhangs at bow and stern tended to make the

175

skaffies trickier to manage in rough weather, but they had more deck room and when beating into places like Wick Bay could turn through the wind very quickly with less likelihood of being caught in stays.

Everywhere else on the east coast the 'fifie' was favoured. It had a very long keel and an almost vertical stem and stern. It was not very safe when running before a heavy sea because the helm was less responsive, but to windward the sharp and deep forefoot kept a good grip. The skaffie was better before the wind because of its fuller after-body. On the wind, though stiff initially, the skaffie lost stability after a certain angle of heel because, unlike the fifie, it carried no ballast. The length and stability of the fifie allowed her to carry a good spread of canvas, and sail faster, but the long keel prevented it from turning quickly and manoeuvring was difficult.

The dipping lug was an unhandy rig, particularly when working into harbour against the wind, but it was simple and by far the most powerful. The great yard was hoisted by a single halyard which was made fast to windward to act as a shroud. There was no standing rigging of any kind, and the mast was lowered by the halyard, so its simplicity, and its efficiency and power when set, out-weighed its clumsiness and inefficiency when tacking. By 1875 fifies and scaffies seventy feet long with a beam of twenty feet and masts fifty-five feet off the deck were being built, some sailing to the south of Ireland and the Isle of Man when the East Anglia season ended. Small fifies about twenty feet long fished near to the coast and were known as 'baldies', which was a corruption of Garibaldi who was in the news at the time the type was introduced.

Both skaffies and fifies were very popular in their respective districts and you can imagine how local feeling must have flared when, in 1879, the daughter of a fifie skipper married at Lossiemouth man who favoured the skaffie, and she brought with her a dowry to help her husband buy a new boat. With the wisdom of Solomon the Lossiemouth boatbuilder William Wood (grandfather of Dr W. J. Lyon Dean, chairman of the Herring Industry Board since 1971) made a half-model of a fishing vessel having the long raked stern of the skaffie and the deep vertical stem of the fifie. The compromise was accepted and *Nonsuch* (as she was called) was an instant success. The type was named the zulu, after the war then being fought in Africa, and was adopted by every port on the Moray Firth, traditional home of the skaffie, and also in many ports that had favoured the fifie.

The zulu was quicker in stays than the fifie and steadier when running than the skaffie; it would lie as close to the wind and drive as easily through a hard sea as the fifie, but it had the larger deck of the skaffie and its twenty-four-foot rudder, hung on five pintles, was better protected in crowded harbours because it was beneath the long sloping stern. Combining the two designs had produced the ideal

True Vine
Fifie type herring drifter, 1905

compromise in terms of functional requirements, but it is also possible
that the builder was thinking of his own problems, for with boats
getting bigger all the time the raked stern of the skaffie was easier to
build, and it was easier to obtain a single piece of timber for the vertical
stem of the fifie.

Zulus were as tough and stout as the men who sailed them. It took
all hands to hoist sail (unless there was a steam capstan). Sheeting in
the huge high-peaked sail in anything of a breeze took the strength of
four men. The masts were like trees, sixty feet high, twenty-three
inches through at deck level tapering to nine inches, and weighing
two tons. The fifty-foot bowsprit could serve as a spare mast and a
thirty-two-foot mizzen boom was run out over the stern. Large supplies
of tallow were carried to grease the chocks on which these huge spars
were slid inboard and outboard as required. The foresail yard was up

41. *Scudding herring from the nets of a Scottish Zulu, probably in Fife around the turn of the century.*

to thirty-two feet long, seven inches through at the sling, and a spare one was carried. The huge foresail, tanned black by successive dippings in cutch, had seven or eight rows of reef points and skippers were as particular about the set of their sails as any yachtsmen.

The boats were massively constructed of two-inch larch planks on oak beams in local yards along the Moray Firth, and were hauled into the water by traction engine. Ballast was beach stones measured by the cart load, the biggest vessels taking forty 'loads' (about forty tons). While fishing, the bilges became very foul with slime and scales and every year they were thoroughly cleaned and new boulders put in. The small cabin, right aft, was shared with the capstan boiler and cooking stove and in summer the combined smell of fish and the humidity made bright metal work such as knives, watches, and even silver coins in pockets, tarnish quickly. Forward of this was a narrow compartment, the full width of the vessel, where the heavy tarred 'messenger' warp to which the nets were attached was coiled down between a 150-gallon water tank and a bunker for two or three tons of coal. About seventy nets were set, forming a fleet one and a half miles long. Floats to support them were made of inflated sheep, goat or dog skins; Buckie men bred dogs specially for the purpose and in Fraserburgh bullock bladders were used. Decks and interiors were heavily tarred and in hot weather every fisherman slept with a knife handy so he could cut his hair out of the soft tar if his head rolled against the planking.

Fishermen were fond of bright colours and the features of their boats were beautifully decorated. Often the black hull had a white cut-water extending all the way aft along the waterline. Blocks and mastheads were usually blue or white, gunwales and the insides of the bulwarks blue or green, hatches and coamings a dark azure blue. Later, varnished hulls were favoured because it was more difficult for builders to cover up bad timber and faulty workmanship. The names and registration letters were picked out in gold-leaf or bright colours.

With such big boats fishing the coast the problem of suitable harbours, never satisfactorily solved, became acute once more. Between 1880 and 1882 Parliament voted seven million pounds to build and improve harbours in England and Wales, two million pounds for Ireland, and only half a million pounds for Scotland. Large sums were spent on a seven-acre harbour at Anstruther, visited by 190 boats a day during the winter, but boats could float only at the top of the tide. On the entire east coast of Scotland there were only four ports with adequate space and sufficient depth for boats to enter or leave at any time – Buckie, Fraserburgh, Peterhead and Aberdeen – and there was an increasing tendency for fishing fleets to become concentrated at fewer ports. By the 1880s Fraserburgh and Peterhead were Scotland's chief herring ports, and Aberdeen was becoming established as a prosperous trawling port.

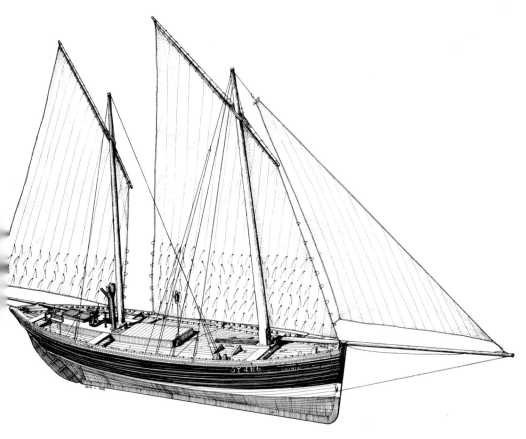

Muirneag
Scottish "zulu" herring drifter with lug rig, c.1880

At the turn of the century the Scottish herring fishery reached a
peak which was maintained until the abrupt collapse caused by the
First World War. Half a million tons of herrings were landed every
year, of which about three-quarters were exported to Germany,
Poland, Latvia, Estonia, Lithuania, Finland, Russia, Rumania and
even USA. Nearly ten per cent was sprinkled with ice and loosely
packed in wooden boxes loaded immediately into specially chartered
ships which sailed for the continent. This system, called 'klondyking'
because it was introduced at the time of the famous gold rush, delivered
thousands of crans of fresh herrings to the delicatessens of the Prussian
empire. The home market for fresh herrings, kippers and bloaters
accounted for about twelve per cent of the catch, and six per cent was
exported as 'reds' to Spain and eastern Mediterranean countries such
as Cyprus, Malta, Turkey, Greece and Egypt.

In terms of human life, the cost of this success was devastating.

Barrels of herrings caught at the main ports of
north-east Scotland in 1863:

Wick	90,000
Peterhead	32,000
Fraserburgh	25,000
Helmsdale	25,000
Lybster	25,000
Cromarty	14,000
Lossiemouth	12,000
Burghead	10,000
Hopeman	10,000
Buckie	8,000
Rosehearty	7,000
Dunbeath	7,000

Every year brought its list of death and disaster, but no calamity was more grievous than the sudden gale which overwhelmed the Berwickshire fishing fleet in autumn 1881. On October 14, the 370 fishermen at Eyemouth weighed up the bright and squally sky and the village barometer which stood ominously low. But the weather had been looking ominous for several days and supplies of costly bait were getting stale, so when a young man scorned the barometer and put to sea forty-two boats with 281 men followed him. The storm broke within a few hours. As the fifty-foot boats raced for the harbour's narrow entrance several were lost within sight of the bewildered and helpless people watching from the shore. The *Harmony*, with six hands, was swamped fifty yards from the entrance. The *Radical*, with a crew of seven, was dashed to pieces on the rocks nearby. The *Press Home*, with six hands, was overcome by the surf. As hundreds of wives, children and old folk watched and prayed another boat, the *Pilgrim*, swept towards the rocky ledge at the west side of the bay with men clinging to the masts and ropes. But she was miraculously picked up by a mighty sea, carried safely over the rocks, and deposited on the other side.

On that terrible day 23 boats and 129 men were lost. One house in every three at Eyemouth lost its breadwinner. The nearby villages also suffered: Cove lost eleven men, Burnmouth twenty-four, Coldingham three, and the Forth ports of Musselburgh and Newhaven lost twenty-four. The 191 men and boys who perished in this single localised storm left 107 widows and 351 children unprovided for. But the survivors had to work to live, and as fragments of wreckage and clothing mingled with seaweed along the shoreline they put to sea again, within a couple of days, though with heavier hearts.

Fishing boats of more than fifteen tons burthen at the Scottish ports of registry in 1883:

Aberdeen	149	
Arbroath	26	
Ardrossan	9	
Ayr	4	
Banff	754	(includes Buckie)
Campbeltown	62	
Dundee	47	
Fraserburgh	703	
Granton	3	
Greenock	30	
Inverness	322	(other Moray Firth ports)
Kirkcaldy	339	(Fife ports)
Kirkwall	108	(Orkney ports)
Leith	165	(Firth of Forth ports)
Lerwick	218	(Shetland ports)
Montrose	118	
Perth	1	
Peterhead	617	
Stornoway	138	(ports of Western Isles)
Stranraer	4	
Wick	549	
Wigtown	2	
Total	4,368	

Petticoat Government.

It was commonly said among Scottish fisherfolk that a woman who was not able and willing to work for a man did not deserve one. Village fish-wives had a hard and ill-paid life that was little better than slavery, but the men could not have done without them. In countless small fishing havens it was the womenfolk – wives, mothers, sisters, daughters – who gathered limpets and mussels for bait and baited the hundreds of hooks on each long-line, mended the herring nets, sold the catch; wives were chancellors of the family exchequer, the force that held families together, and in the days when boats were launched from the the beach wives would 'kilt their coats' (petticoats) and carry their husbands aboard on their shoulders so they started the voyage dry. As late as the 1930s it was considered a real gamble for a 'loonie' (a boy in the fishing community) to take a town 'quinney' (girl) as a wife, for without a woman who knew the business a fisherman was as helpless as a boat without a sail.

Fishing in Scotland was entirely a family affair. Daughters were brought up to load net-mending needles with twine before they were seven, and turn the heel of a sock without the aid of a knitting pattern before they were ten. They twisted 'snoods' from ten strands of horse-hair for the long-lines and helped their mothers bait the hooks and prepare their father's 'kit' or lunch-box, which was like a small half-keg that he carried over his shoulder on a rope. Children helped make oilskins by painting cotton with ochre then rubbing in layer upon layer of linseed oil with a pad of cloth. At harvest time they collected chaff to fill mattresses used by their fathers and brothers at sea, and made them blanket bags covered with an old sheet using big stitches so they could be unpicked for washing. There was also the job of collecting blades of grass with which to separate the coils and snoods of the long-lines as they were laid ready in the tray. A fisherman's daughter was as busy as her mother.

Ashore, the men did very little but look after their boats, spending their time sitting on the foreshore smoking short clay pipes that were not considered broken in until they were stained mahogany brown by tobacco juice. As long as their wives had a penny in their purse there was no need to go to sea, but once it became necessary their lives were harsh and dangerous in the extreme.

Before railways the only way to find a market for fish was to hawk it from door to door or carry it to a town or city. At Fisherrow and Musselburgh, when the boats came in late, women ran their fish baskets five miles to Edinburgh in relays. There is one account of three fish-wives covering twenty-seven miles from Dunbar to Edinburgh, each with two hundred pounds of fresh herrings on her back, in five hours. The fish-wives at Newhaven, then a picturesque fishing port just out of Edinburgh, became famous when George IV told Sir Walter Scott they were the handsomest women he had ever seen. When Queen Victoria admired them they were modelled in cardboard, made up as whisky bottles, given to children as dolls, and their brightly striped petticoats hitched up on either hip to make pads for carrying baskets were widely illustrated as the costumery of fisherfolk. Carrying more than a hundredweight of fish in a basket supported by a leather band that fitted over her muslin cap, the industrious fish-wife sold her fish from door to door for a few pence at a time and earned a well-deserved reputation for haggling which she could swiftly terminate with the remark, 'Fish are no' fish the day, they're just men's lives.'

When railways in effect brought markets to the quay-sides the tradition of the fish-wife's involvement was continued in quite a new direction. As Scottish herring boats became bigger and more sea-worthy, and began to follow the shoals to different parts of the country, Scots fisher 'lassies' followed the fleets to do the gutting and packing. For the East Anglia season which started in September whole towns up-rooted and travelled south to Yarmouth and Lowestoft in special trains. Miss Mary Murray, a school-headmistress who has retired to live in Anstruther, Fife, where she was brought up as a member of a large fishing family, remembers her grandmother describing a voyage to the Shetlands. The voyage was made in an open boat, in about 1870, when she was aged about twelve, 'There was no place to sleep, you put your head on a float made of a pig's bladder and in hot weather it stuck to your cheek when you woke up. Each boat in the small convoy that sailed for Vatersay had a small girl whose job was to make the bannocks (scones) of oatmeal, organise the beer and buttermilk for the crew of her boat, and in general to attend to their creature comforts. While the boat was at sea she did her chores and looked for water which she found by digging in a damp patch in the shingle.'

The fleet-following lassies, known universally in the south as 'Scotties', were fishermen's wives, daughters, sisters and widows, and sometimes their grandmothers. One gutting crew comprised a great-grandmother, a grandmother, and a great-grand-daughter whose ages ranged from seventeen to seventy-five. In the Shetlands they lived in rough huts with few comforts, sleeping three to a bed. There was no furniture except their own wooden chests, or 'kists', which served as trunks, chairs and tables. In Yarmouth they lived in lodgings found for them by curers, and sometimes in prefabricated huts erected among the mountains of barrels in the pickling plots on the denes. Their accommodation was always scrupulously clean and you would often see a jar with a few wild flowers decorating a window-sill. In crews of three they gutted and packed herrings non-stop until the work was done, even if it continued long into the night by the light of flickering flares. To protect their fingers against knife-cuts and rough salt that burned into their wet hands they bound their fingers with thick linen bandages called clooties which were tied with fingers and teeth and every night washed and pegged out to dry.

The arrival of the lassies at Great Yarmouth in September 1903 was described in *The Gorleston Times*: 'It is interesting to see the newly arrived Scotch specials disgorging their squadrons of red-haired, ruddy-cheeked passengers with countenances stolid and innocent of wonder. A hard life has chiselled a look of serious unsentimentality on their faces. And the way a brawny lassie assists the would-be friendly porter to trans-ship her boxed-up belongings exhibits more brawn and sinew than feminine feebleness. Further, the dare-you-touch, nothing-will-flurry-me manner of sitting on that same trunk until its turn comes to be jerked upon the already heaped-up barrow strikes you as being something new and characteristic, not to say useful, in a muscular lass. The lassies quite alter the character of our environment. And what fine specimens of womanhood they are. After work, or when waiting for the boats to come in, they roam around, clean as the proverbial new pin. Their faces are aglow with health, their hair, innocent of covering, neatly and fashionably done, and with good warm clothing in strong contrast to the shoddy dress of their southern sisters, they present the very embodiment of strength and comeliness. They are happy, too, despite their arduous, unpleasant work – ever singing and busily knitting as they take their walks.'

At work, the picture they presented was markedly different, as *Toilers of the Deep* described in 1905: 'The old town rumbles and shakes with them, for most are booted to the knee with mighty Wellingtons an inch thick in the sole and so heavy that they drag and rasp along the echoing pavements. Honest work is their business and homespun their dress. They have pretty hair, of all shades of red and auburn and brown, but a Paisley shawl fastened tightly under the chin hides all such worldliness. Their arms are naked to the winds, red and rough. Their fingers are nearly all in bandages, for their trade is with the knife and the blade is forever slipping across a wriggling herring and bringing disaster and gore.'

Standing on stones in a mingled mass of mud, scales, slime and herring guts and unprotected from the weather, the lassies could carry a barrel on each hip and gut twenty thousand herrings in a long day for which, early this century, they received eleven shillings a week with accommodation and travel expenses paid. It was a bloody business and one writer in *Blackwoods Magazine,* about 1870 saw it as the fish-wife's revenge: 'When gutting is at its height their hands, their necks, their busts, their "dreadful faces throng'd, and fiery arms", their every bit about them, fore and aft, are spotted and besprinkled with little scarlet clots of gills and guts. A bob down to slice a herring and a bob up to throw it into the basket and the job is done. These ruthless widows seize upon the dead herrings with such a fierceness as almost to denote revenge for their husbands' deaths – victims of the herring lottery – and the widows scatter about them the gills and guts as if they had no bowels of compassion.'

It was a hard life but many women in Scotland still remember it well and always with affection – 'It was *great* fun.' Each group had a cheer-leader and often a whole yard of perhaps twenty crews of gutters and packers broke into song, usually a hymn. There was no work on Sundays, partly because the Scots were keen Church goers but mainly because it was widely believed that one day off in seven gave the herring shoals a breathing space during which they regrouped. A lassie courted young and married a fisherman, usually from her own village. Although very closely knit the Scottish fishing communities had a wide outlook because they had seen a lot of the world. Considering the sanctimony of the age they also had a broad morality. Young men and women far from home for long periods were quick to take advantage of each other's company, and it was said that a young man needed to get a girl pregnant before he proposed to ensure that she was capable of bearing children. If he succeeded he always married her, and if the number of pregnant brides was remarkably large, divorces and illegitimate children were practically unknown. Nobody looked askance as long as the couple settled down afterwards, and it is likely that the casual attitude towards pre-marital sex was another aspect of the fisherman's fatalism.

Scots fisherfolk set a high value on a good home. Despite the roistering that went on when a crew got a few whiskies under their jerseys their basic values were down to earth. As fishermen they were more professional and as folk they were often more solid than their counterparts south of the border. Fisherlads had a reputation for industry and integrity as apprentices in other trades. Family priorities were first a good boat, well looked after; second, a house with good furniture; and at least one child at university. Coastal villages in Scotland today are filled with returned exiles, like Miss Murray the sons or daughters of fishermen who have led professional careers and returned to retire. It is not in human nature to return to a place where you were unhappy.

Sudden death was a daily factor in the lives of every fishing family. People knew sorrow but they knew how to face it and there was little weeping or wailing, at least not in public. When condolences were offered friends and neighbours were plain about it – 'Your man's away, what an awfu' job.' People did not let each other sink. Community events were held to raise money for destitute families. Widows invariably turned their hands to making oilskins, knitting, or mending nets. When things got bad there was often a basket of food left on a poor woman's doorstep. A bag of potatoes cost almost nothing and there was always fish available on the quay. Grocers in some ports still have outstanding bills that they never pressed, but it could turn to advantage when the sons of the family grew up and provisioned their own fishing boats – 'We'll get our food at Fowlers, he brought us up.'

Early in the last century a typical fisherman's cottage had a single room with a high roof and small dark windows. Among the rafters, black with peat-smoke, were hung fishing nets, dried fish, onions, dried herbs for home-made medicines, bunches of sedge with which to clean clay pipes, and rush-wicks for lamps that burned oil made from crushed fish livers. A tinder box made from a cow's horn filled with charred linen was used by the fisherman to light his pipe. There was a large box bed in a recess and the bumpy earth floor was sprinkled with sand after the lines had been baited; the evening sport of the children was catching sand-hoppers jumping away from the fire.

At the end of the last century a young man moving with his bride into their first home expected a 'single end' which was one large room in a four-roomed house. Larger families had two rooms in a house which had a big wash-house in the yard, cellars, and an attic for drying nets. A woman's status symbol was a 'braw press' or tall cupboard beside the hearth – the door always left just ajar – in which were arranged china gifts that had been brought back from different herring ports by her menfolk. Seaboot stockings hung to dry over the cast-iron cooking range where salted herrings, called 'bufters', were roasted over a glowing fire on the 'brander' and eaten with the fingers. When the boats went away to Yarmouth a telegram was sent with news of their arrival and Mary Murray recalled running round the town taking 'the word' to the home of each man on board. Money was always divided in the skipper's house and wrapped in a red handkerchief, small amounts knotted in each corner, which Mary Murray remembers taking round at the age of five. Crews tended to stick together, the arrangements being renewed by word of mouth at the end of every

42. A Scottish fish-wife who travelled out into the countryside of Moray-shire by train to hawk fish around farms and outlying parishes.

43

"A FISHER SCENE, BUCHANHAVEN, PETERHEAD.

44

season. The whole town rang with the scandal of it if a man dropped out of a crew and let his side down.

At sea a man protected his upper half with a big buttonless oilskin frock called a dopper, and a pair of half-mast oilskin trousers, with a leather seat for comfort during long periods of rowing: the legs were short, falling just below the top of the boots, because fishermen were so often wading up to their knees in herrings. Under these he wore a pure-wool flannel work-shirt, and a pair of kersey breeks made of thick, blanket-weight woollen cloth which had a flap in front with a button at each side and were taped just below the knee. If he had a bad back a thick quilt was sewn inside his shirt to cover his kidneys. Navy-blue guernseys were knitted in elaborate patterns, the front and back in one piece so they were less likely to come apart. Stitches were counted in scores – 'Dad takes nine score and ten' – and the women and daughters became very proficient at mental arithmetic because the complicated patterns were never written down and the sums had to be done in their heads. Thigh-length leather boots with heavy iron heels and inch-thick soles studded with nails were made to order for the cost of a week's earnings and lasted seven or eight years; when the leather became stiff the men walked with a comic gait. Immense sou'westers with the long brim turned to the front in sunshine and to the rear when it was raining were the commonest head-gear, but seal-skin caps were also common and until the 1850s a tall silk hat similar to a topper was worn at an angle like the lowered mast of a drifter and called a rakie stop. On Sundays, out came a dapper suit of blue serge or pilot-cloth with a double-breasted jacket, waistcoat, and a cheesecutter hat.

Although they were conservative, fisherfolk were also emotional and easily moved by any form of organised Christianity. Religious movements such as the Salvation Army and the Brethren swept whole communities into their folds in the latter part of the last century. But religion went only skin deep compared with the fisherman's instinctive sense of the supernatural. Superstition governed almost every aspect of his life. Up to the First World War lonely old women in the Scottish fishing villages often had reputations for possessing supernatural powers, and a man would not put to sea if he met one of them while on his way to his boat. Nor would he put to sea if he met a red-haired person, someone with flat feet, and a four-legged beast was unluckiest

43. *A fishing family mending drift-nets and preparing to sell fish at Buchanhaven, Peterhead, around the turn of the century.*
44. *Scottish fishermen cutting each others' hair in the bows of a drifter: early wooden vessels had tarred hulls and fishermen slept with a knife handy to cut their hair out of the tar.*

of all – even a mention of a pig or a rabbit in the hearing of a fisherman tempted fate. These superstitions are common among all fishermen and persist even in the big freezer-trawlers of today; their origin may lie in the fact that pork and rabbit were the main competitors of fish as a food item. In Grimsby fishing operations were once brought to a temporary halt when some boys drew the outline of a pig in the misted window of a pub crowded with fishermen, none of whom would put to sea for twenty-four hours on account of it.

Whistling at sea brought gales. A Scottish fish-wife never did her washing on a Monday, when her husband put to sea, in case she washed him away. A minister was welcome at a boat-launching ceremony but was seldom seen at the dock-side and was never spoken of afloat except in vague terms as 'the man in the black coat'. It was bad luck for a woman to cut her hair after sunset if her husband was at sea, or to blow on hot oat-cakes. Certain surnames such as Ross and White were unlucky so 'tee-names' or nick-names were handed down through generations – Langlegs, Shavie, Pinchie, Howdie Doddlies, Old Bar, Fly, Tarry, Fleukie, Haddie, and many others.

Whisky was drunk to seal bargains, bless the nets, and keep out the cold. On the whole, Scotsmen visiting the English ports had a reputation for integrity, thrift and sobriety and it was rare for them to drink before the day's work was done, but when they did drink it was with abandon. One Scottish skipper whose lugger was in Great Yarmouth for the weekend gave the cook-boy money to lay in food and drink; he came back with half a dozen bottles of whisky and a penny loaf. For a moment the skipper gazed admiringly, but then he complained: 'Heaven forgive you, boy! Whatever shall we do wi' all this bread?'

But the true role played by whisky in the fortunes and misfortunes of Scottish fishermen is hard to evaluate. After the Wick and Peterhead disaster in 1848 Captain Washington condemned the practice of curers who made part of their payments to the herring boats in whisky. On the other hand Mary Murray remembers there was at least one bottle of whisky in every boat 'to help them see in the dark – one boat had the same bottle for thirty-six years.' It would seem that about two-thirds were abstemious but the rest drank a lot. Those who did not attend the meeting at the Temperance Lodge every week got paralytic-ally drunk and were taken home in barrows.

The truest picture of the typical Scottish herring fisherman was described by J. R. Bagshawe in *The Wooden Ships of Whitby*: 'After a

---

45. *Scottish fisher lassies gutting and packing herrings on the curing plots at Great Yarmouth.*

46. *Herrings being packed for export in a Scottish fishing port, probably in Fife.*

hard night at sea and a busy morning getting rid of his fish he may sit for a few minutes on deck with a mug of tea and some bread and butter and herrings in his grimy fist – there is no time to wash and water is scarce. The signs of fatigue are plain on his face, which is begrimed with galley smoke, channelled by perspiration, and plain in his heavy, stumbling feet and bent back. Twenty minutes at most he will allow himself. Then, with his blackened clay stuffed into his mouth, he wearily starts afresh.'

THINK O' THE STORM, WOMAN!

Life in a small market town with a harbour and fishing fleet on the north-east coast of Scotland in the 1880s, described by A. N. Meldrum in a 1950 issue of The Countryman.

Although there was not a single fish-shop in the place we had a regular supply of fish brought to the door by fish-wives who had their regular customers and could deal effectually with anyone trying to poach. I remember the prices well because it was fun to be at the door when Meggie, or Cauld Annie, came with her basket. It was always assumed that the fish would be bought. Indeed, it would have been impolitic to refuse.

For the most part the fishing boats brought in haddock, whiting, hake, cod, plaice and herring. For haddocks of frying size, to satisfy the hunger of one hungry man, the price was fourpence to sixpence the half dozen. Whiting was cheap and little valued and could be had at threepence for six. Small cod was rarely bought. Large cod was in the main dried on the clean beach shingle, being left there for three weeks after being soaked in brine, to become the 'hard fish' of village shops, on sale at anything from tenpence to 1s. 2d. Soaked and boiled, it gave some variety to farmhouse and bothy food. A good-sized skate could be had for sixpence or sevenpence and in my mother's day was much sought after at spring-cleaning time since the water in which it was boiled was kept as a foundation for whitewash or yellow ochre when the kitchen back-regions were done out.

The herring season was the town's busiest time. I have often seen a fisher woman filling a pillow-case at the baker's with tea-bread and penny things – buns, cookies, cheese-cakes, queen-cakes, jam sandwiches and such like – for the men in the boats. Herring of the best quality, fresh from the nets, could always be had for three or four a penny later in the afternoon from a hawker with a barrow. If one bought more cheaply at the shore-head, from a fisherman carrying a string of live crabs holding on to each other by the main claw, he detached the purchase by spitting into its eyes.

Yellowfish, or finnan haddocks, were best cured in a small way by the fishwife herself. They were invariably sold in threes, tied together at the lug by a bit of rush, 3d. to 5d. a bunch. Any higher figure was contested, but the fisherwife generally carried the day with her shrill, 'Eh woman! The wather! Think o' the storm!'

The Coble Coast.

Between the River Tweed and the mouth of the Humber the whole
coast was dotted with fishing communities. Fish was carried inland by
pack-horse from the rock-bound harbours of Northumberland and the
steep beaches of Durham. In Yorkshire fishermen colonised the spurs of
cliffs like Robin Hood's Bay, where a steep street twisted down to an
open beach, and narrow ravines like Staithes where four hundred
fishermen kept eighty boats in a tiny creek. South of the border the
beamy and double-ended Scottish herring boat gave way to a different
kind of vessel, which was in universal use on this difficult and exposed
coast where the majority of landing places used by fishermen did not
have so much as a breakwater to protect them from the fierce North
Sea surf. This was an open boat with broad strakes, a high battle-axe
bow made to chop through curling breakers, and a sheer ending in
pronounced tumblehome providing a unique prettiness of line and
serving the purpose of keeping the water out when heeled. It was
called a coble.

The forward part of the coble was all Viking longboat, the great bow
built to shoulder through any weather, enabling the boat to be launched
from an open beach at any time. A thrash of twenty or thirty miles to
windward was an every-day task on this coast, and the boats were built
for it. About halfway along, the deep forefoot flattened out and the
keel disappeared. What took its place was a flat bottom on which were
fixed twin keels, the aft ends curved like the runners of a sledge. This
enabled the boat to be hauled stern-first out of the breakers and up to
safety beyond the high-tide mark by teams of horses or steam engines.
For centuries a distinctive feature of England's north-east coast has
been the sight of dozens of brightly painted cobles lying on the beach
landings like herrings on strings, and although they are now launched

47. *Whitby fisher-girl baiting long-lines with mussels and limpets which
she has collected on the foreshore, c.1880.*

48. *Whitby fishermen c.1880 gossiping on the quay displaying a variety of headgear, sou'westers at all angles, and the heavy leather fishing boots that carried so many men to the bottom of the sea.*

Enterprise
South Shields sailing coble

by rusty tractors, and the number of boats is greatly reduced, the scene is little changed.

Modern cobles are decked-in forward and powered by diesels, but until the First World War they all had low, raked masts with lug rig. The rudder, controlled by a long tiller, was very deep so it acted as a

---

49. All the family is involved when it is time to tar their coble at Whitby, c.1880. Long-lines, oars, and other gear is heaped up in the foreground. The fish-wife is setting off to collect bait for the lines.
50. Crew of a Yorkshire coble at Whitby c.1880; the girl carries a long-line on a tray.

kind of centreboard to give the stern of the vessel grip on the water. The name coble (sometimes spelt cobble but always pronounced like 'noble') is thought to be Elizabethan in origin but the hand of the Norseman is evident in the craft's massive construction. Running before the wind was its weakest point: it was inclined to go too fast, lose steerage, and broach. There was no place to make the mainsheet fast so that in a squall it could be quickly eased to spill the wind; there was always a man ready to drop the sail if the boat 'ran away'.

If the coble could be launched in any weather at the expense of only a few wet legs, getting ashore through the breakers was something else. In moderate weather the coble came up into the wind just offshore, sail and mast were lowered, the long rudder was unshipped. Then the three or four hands, standing at long sculls, backed the boat towards the beach in a cloud of spray until she touched. At Flamborough Head the cobles were hauled up the steep rocky landings by steam engines driving winches, after the catches had been loaded into donkey paniers. At Filey they were run on to old-fashioned boat-trailers with huge cartwheels, which were pulled up the sands by work-horses.

After one wild storm in the 1870s all the Filey cobles had landed but one. The *Filey Post* described how hundreds of people waited on the shore for its safe arrival. Houses facing the sea put candles in their windows as guidance and a fire was lit on the beach to warn the crew – William Hunter, his son, and Matthew Jenkinson – not to land until the lifeboat was launched. 'Thinking the fire was a signal for him to land at that spot, Hunter turned his boat's head to the breakers and made all haste. One like a huge cliff was seen approaching which filled them with terror and dismay. It broke upon them with a crash but the little buoyant coble instantly rose, shaking as if preparing for another attack. Now they pulled and strained every nerve during a momentary lull. At this moment, as the lifeboat was about to move, the tiny fragile craft was seen making its way through the foaming waves. A shout was raised and some intrepid fishermen rushed into the water, holding fast by each other, and seizing hold of the boat dragged it up the beach.'

For line-fishing and potting crabs and lobsters between December and May fishermen used small cobles about thirty feet long which carried three men and were handy and nimble in rough waters but could be more easily beached. In June the larger five-men cobles

---

51. *Boys helping to unload a Yorkshire coble on the foreshore at Scarborough on a calm day.*
52. *Wrecks were a common occurrence at haven entrances when fishermen misjudged wind or tide then were unable to escape seawards; often men drowned within sight of families watching on the shore.*

called ploshers were launched for the herring season. Shooting the long-lines required confidence and a steady stance in the boat. The helmsman luffed to control speed and had to be careful not to foul the lines with the deep rudder. Hauling was done under oars, bow-first in a wind and stern-first in a calm. After a long back-breaking row home in smooth weather the fishermen's wives met them at the dock-side with Fuller's Earth to ease the soreness under their arms. Dogfish did terrible damage to the lines, raiding hooks and killing senselessly like foxes in a hen-run. If he caught a dogfish a coble fisherman tied a cork to its tail so it floated until it died – 'Nowt's over bad for them owd dogs!'

Fish-wives were known as 'flithers women' because they scoured the cliffs and low-tide rocks for flithers (limpets) and mussels, and dug worms from the sands, for bait. This was always in short supply; mussels and herrings were bought from Scotland, whelks from Grimsby, and lampreys were imported in barrels from inland towns such as York, but each time the train stopped the barrels had to be stirred to keep the water in motion as they live only in running water. Sometimes bait had to be bought at great cost from Morecambe and Holland. Filey women who went 'flitherin' to Scarborough were provided free of charge with a special truck for handling their mawns (baskets) at the railway station and they travelled in a reserved carriage at reduced rates; it was said the difference was made up to the railway company by a benevolent local lady.

Through most of the last century the fishermen at Filey were a unique community, keeping themselves apart not only from other inhabitants who opened lodging houses but also from other fishermen. They lived in the old part of the town and did not mix. Their thirty-five small cobles were launched off the beach, and they had about the same number of yawls which operated from Scarborough. These were about fifty feet long with a beam of seventeen feet, and had two masts with lug rig. Many were partly owned by 'gentleman speculators' such as harbourmasters, other boatmen, local traders, and relatives of the skipper. In the 'spring fishing' from January to mid-summer they went long-lining. Every Monday the yawl fishermen left Filey on the 9 a.m. train for Scarborough, accompanied by women carrying the week's provisions in a tin grub box on their heads. They sailed about thirty miles out, shot long-lines from the yawl and hauled them from a small coble carried on deck. As a rule they returned home for the Sabbath, landing their catches on the sands where their womenfolk met them with carts and the skipper came ashore to auction his own catch. The second half of the year was spent drifting for herrings. The Reverend George Shaw went to sea with Filey fishermen in a yawl in 1866: 'After shooting 120 nets in the water now comes the summons to supper. This consists of beef and fish and very substantial it is. You sit down upon the deck. A piece of broiled beef, smoking hot, is brought up and one man cuts it up and hands you a piece big enough for half a dozen meals. With this is served bread – home made and worth the name of bread – cabbage, potatoes or carrots. Should there be a sea you must cut your meal up at once or chase a rolling potato across the deck. After the meat come the dumplings which are served with raisins and the treacle can is handed round. Then – a dish o' tea.'

The biggest fishing ports on the coble coast were Whitby and Scarborough. The latter's indifferent port was useless at low tide and larger boats had to lie outside and discharge into small boats or carts, but it was the first North Sea trawling port and by 1868 it had a fleet of seventy-two beam-trawl smacks which caused serious ill-feeling among the yawl men. Despite its small size, the harbour was crowded with sailing drifters during the herring season – 50 Penzance boats, 200 from East Anglia and 300–400 from Scotland. The harbour-master entered in his logbook on October 5, 1905: 'Terrific gale with high sea, both harbours full to entrance.'

By the end of the nineteenth century steam trawlers from the Humber fished the east coast intensively and the coble fishermen were the first to feel the effects of it. Between 1893 and 1913 the quantities of line-caught fish caught by cobles dropped to one third, and as a result the number of cobles was halved. North Shields and Hartlepool became flourishing market centres because steam-powered vessels could obtain cheap coal there, but in the small ports fishermen followed the traditional work of line-fishing only to obtain bait for potting. Salmon-netting, cockle-raking and certain specialised fishing methods continued, such as catching turbot in 'brat nets' (anchored nets with large mesh) but the younger generation was lured away by coal-mining and other more lucrative work while women found easier and better-paid jobs than flithering. The fishermen who remained found they could turn a penny or two by filling their cobles with summer day-trippers and holidaymakers.

Today the coble is still characteristic of the small fishing boats found on the coast of Yorkshire, Durham and Northumberland. They are engine driven and launched from trailers by rusty tractors, and when there is a bit of a sea running you can still sense some of the excitement and anxiety of getting a coble away to sea in the days of sail and oar.

Spat, Sprat and Bawley.

The Thames estuary was the world of a unique type of fisherman, a man able to turn his hand to anything from oyster dredging to yachting. In these shallow, muddy waters where the sea sluices banks of ooze and washes prairies of hard sand, and where long fingers of tide creep inland along the Essex, Suffolk and north Kent shores, were found the shrimpers, the cockle-rakers, the whitebait-netters, the watch-boat on guard against oyster poachers, and the anchored stow-boater with his enormous net that scooped up tons of sprats at a time.

As the trawling smacks moved further out into the North Sea after 1815, and Barking began to decline as a port in the 1860s, those who remained to fish their home waters greatly diversified their techniques according to the time of year and what was fetching the best prices at Billingsgate. Shipwrecks occurred frequently in those busy and difficult waters and every fisherman was also a salvage expert, his boat fitted with crowbars, grapnels, and axes for pillaging wrecks and saving lives. Fishermen did not get rich: their simple and hard-working lives were scantily rewarded, but if they did not have cash in their pockets they were in most respects better off than agricultural labourers.

Sprat-fishing with a stow-net was a grim task because it took place in the middle of winter. It was an ancient method of fishing: in Elizabethan times at least a hundred 'stall' boats fished in the Thames, so named because boats and nets were anchored or stalled against the tide. Over the centuries the name became corrupted but the method did not change much. By the middle of the last century a typical 'stow' boat was a cutter about thirty or forty feet long. It rode to a heavy-duty anchor with an immense tapered net slung beneath the keel and held against the tide with a rope to the anchor. The mouth of the net, kept open by horizontal spars called baulks, was nearly thirty feet high and twenty feet wide. The net was ninety feet long, and the main danger was that it could fill with a shoal of sprats in a matter of minutes and become an unrecoverable dead-weight.

In the 1860s stow-boat fishermen fought each other for berths at Billingsgate, discharging up to 130 tons a day. In the 1870s an estimated 600 boats and 2,600 men fished for sprats from both shores of the Thames Estuary. Often there was a glut, dozens of boats catching enormous quantities, and the fish were sold at a pittance for manure. Small coasting vessels also stow-netted for sprats, filling their holds then sitting on the mud, giving off a foul reek, until the fish had rotted sufficiently to be sold to farmers who collected it in carts at low tide. At Brightlingsea sprats were cured by fishermen in smoke-houses and sent to London, or pickled in barrels with bay-leaves, salt and spices, and exported to Holland and the Baltic where they were tinned as sardines and sometimes found their way back to Britain.

Whitebait, the small-fry of herrings and sprats, were caught in the same way but in smaller, finer nets deployed from smaller boats in shallow waters. It became a popular dish in the 1850s.

In shallow water with a soft muddy bottom large numbers of shrimps were caught in very fine nets. On the Humber sands and the Lincolnshire coast, and at Blackpool and Southport, trawl-like nets on T-shaped frames nine feet wide were shoved along by men leaning on the long handles with their chests and carrying baskets on their backs for the catch. Where the bottom was firm enough shrimp trawls were attached to the axles of one-horse carts driven through the water. The succulent pink shrimp, often called a prawn but in fact a different species, was sold for potting and it found a big market on seaside stalls. Throughout the last century shrimping was also a widespread cottage industry on the Thames; ports such as Leigh and Gravesend each had fleets of several dozen boats. These were shallow-draft, heavily built cutters with vertical stems and transom sterns which were known as bawleys. At first the shrimps were boiled in sheds ashore, then the Leigh bawley *Surprise* fitted a boiler on board and by 1850 there were 100 boats sailing from the port and boiling their catches at sea. During this century their numbers declined and today only three boats fish part-time from Leigh.

An average shrimper could make two or three pounds a week at the turn of the century, which was shared by the crew of two. A bag-shaped trawl net was cast on the edge of a bank of grey ooze; thirty or forty pounds of shrimps was a good haul. They were tipped straight into a vat of fresh water kept simmering over a coal fire and boiled for about five minutes. Shrimpers made a picturesque sight as they came home at sunset, pushing up the creek with the tide. But in winter it was dreary work and not without its hazards. In 1901 a Leigh fisherman told *Toilers of the Deep*: 'I once worked aboard a shrimper which went on a mud bank and broke up. Every man knew there was only one thing left for him this side of death, and that was to sink slowly in the black ooze of the bank till it covered his head. Me and my mate could swim,

Boiling shrimps aboard an Essex bawley.

so we swam to a buoy and climbed on it, and from there we watched our mates try to crawl along the ooze, and go down one by one.' Another bawley run down by a steamer came to rest with fifteen feet of its mast jutting above the water. One of the two men who survived the collision clung to the mast-head in winter darkness for thirteen hours before he was rescued, the tide creeping up to his arm-pits before it receded.

Flat-fish caught by bawleys that went line-fishing were often kept alive in pits, 120 feet long, which were dug in the marshes, puddled with clay, and protected from high tides by embankments. Turbot were put in with corks trailing from their tails on pieces of string. The fish were taken out when the fishermen were ice-bound in winter and fetched high prices at Billingsgate. One ancient method of catching flat-fish was banding. Old men earned a penny a hundred for making wooden hooks from blackthorn which were baited with lug-worm and set on mile-long lines pegged out on Foulness sands at low tide: the wood gave the bait just enough buoyancy to lift it off the bottom and it was not unusual to find a flounder on every one.

Silting caused Leigh to decline, but the extensive sands which were created provided huge resources of cockles. Boats went out on a falling tide, the men sounding over the side in shallow water and listening for the crunch of cockles under the oar. Ensuring that the bows of the boat pointed seawards as the tide went out from under it (in case the weather blew up later) the men then waded to the drying sands and raked cockles, boiling them ashore in sheds. Until 1967, when the White Fish Authority invented a suction dredge, all cockling was done by hand using garden rakes.

But it was for its oysters that the estuary of the Thames had been best known – since the Roman governor Julius Agricola began exporting them to Rome in A.D. 78 and initiated a trade that lasted for three centuries. Even the crushed shells were utilised by the Romans, as cement, tooth-powder, and a cure of chilblains. Since earliest times oysters had also been harvested from Holy Loch near Glasgow and Prestonpans near Edinburgh, from Chichester and Poole Harbours and the Hamble on the south coast, the Exe and the Fal in the west, the Mumbles in Wales, and The Wash and north Norfolk in the east. At Poole a breakwater was constructed in the harbour from the shells of oysters which were opened for pickling and sent in barrels to London, Holland, the West Indies and Spain in the time of James I. But the shallow, warm and silty waters at the mouth of the Thames were ideal for growing oysters, and it was only these (although supplies were augmented with young oysters from other regions) that were known as 'true natives.'

Until early in the last century oysters were a staple food of the poor and were hawked round the back streets in great quantities. Usually they were 'scalloped' (cooked) to prevent their going bad, but then they had little of the piquant flavour so popular with Roman senators and modern expense-account diners; Dr Johnson referred to scalloped oysters as 'children's ears in sawdust'. Prices began to rise when grounds such as Prestonpans declined due to pollution and over-fishing, and demand from the new industrial cities exceeded supply. In the 1840s you could buy fresh oysters for three a penny, in the 1860s two for threepence, and by 1900 they were fourpence each. From the middle of the last century until prices became prohibitive the oyster supper was a fashionable and mouth-watering occasion, like this one in the Midlands held just before Christmas in 1863: 'In the centre of the table, covered with a clean white cloth up to the top hoop, stands the barrel of oysters, a kindly remembrance from a friend. Each gentleman at table finds an oyster knife and a clean coarse towel by the side of his place, and he is expected to open oysters for himself and the lady seated by his side, unless she is wise enough to open them for herself. By the side of every plate is the *panis ostrearius*, the oyster-loaf made and baked for the occasion, and all down the centre of the table, interspersed with cases of bright holly and evergreens, are plates filled with pots of butter, or lemons cut in half, and as many pepper and

vinegar castors as the establishment can furnish. As the attendance of servants at such gatherings is usually dispensed with, bottled Bass or Guinness, or any equally unsophisticated pale ale or porter, is liberally provided.'

Disaster of a kind occurred on November 10, 1902 when the guests at two mayoral banquets held simultaneously at Winchester and Southampton became ill with typhoid and four died. Investigation showed that oysters at both banquets had been supplied from the same beds at Emsworth, which lay just below the town's sewer outfall. The Fishmongers' Company took samples of oysters from beds all round the country and found that many were 'utterly unfit' or 'grossly polluted' by sewage. Oysters from these grounds, and other shellfish such as mussels in the Exe and Teign, and Southend cockles were banned. River Colne oysters were not affected because Ipswich sewage was chemically treated. Whitstable was in trouble because its tidal pits where oysters were stored before being sent to market were polluted, but its six square miles of oyster beds offshore were safe. Systematic bacteriological sampling of oysters prevented further serious outbreaks but many oyster and mussel beds never again received a clean bill of health. Until the 1950s oysters were purified in big pits in the marshes where the water was known to be clean; now it is done by washing them in sea-water which has been purified by passing under ultra-violet light.

Oyster-fishing is the only branch of the fisheries in which the fisher-man is partly farmer as well as hunter. Methods of farming oysters began in the Thames early last century when 'brood' or tiny one-year-old oysters were brought from other regions and sowed like seeds on the beds to fatten for two or three years before they were harvested; sometimes larger oysters were laid, and fattened for only a year. They came in large quantities by coaster from Britanny, Falmouth, and from the River Tagus in Lisbon. Portuguese oysters, with a distinctive long and crinkly shell, were cheaper than native flat oysters and sold mainly in seaside stalls (a steel factory so polluted the Tagus in 1965 that now Pacific oysters, which look similar but mature in two years instead of three or four, are extensively cultivated in the Thames estuary). The Welsh beds were practically denuded of oysters which were re-laid in the Thames. There was also an interchange between the north and south shores, Suffolk and Essex oysters being fattened in Kent where, in the course of a year, they lost the greenish tinge that

53. Cockle-rakers working on the sands off Leigh-on-Sea, Essex.
54. A stow-boat working in the mouth of the Thames Estuary retrieves its net and a lantern is held aloft to see whether it contains a good haul of sprats.

211

was caused by a growth of summer seaweed and led inexperienced diners to assume they were 'off'. A certain uneasy rivalry between the Essex/Suffolk men and the men of Kent had always existed. In 1724 a Captain Evans, then mayor of Queensborough and an M.P., led 100 smacks in a raid on the Essex beds and 12,000 bushels (500 oysters per bushel) were taken. Three years later £20,000 damages were awarded against the Medway fishermen who took twenty years to pay it off.

Oysters are harvested between September and April (when there is an 'r' in the month); at other times they are releasing spawn, or 'spat'. At Whitstable at the turn of the century there were about a hundred oyster smacks, each about forty feet long and cutter rigged. Most were owned by their skippers, and they were hired by the day by the foremen of the two principal oyster companies. Working across the tide, these bawleys and yawls each worked a 'fleet' of five or six iron dredges like small, heavy-duty trawls. The warp of each one was coiled on deck and buoyed so if the dredge came fast it could be released and recovered later. The dredges were hauled frequently, small oysters being thrown back overboard and the big ones separated from the small soles, eels, squid, crab and seaweed that was also collected. Sail continued in general use at Whitstable until the Second World War because engines were too powerful and tended to tear up the beds; on the oyster grounds at Falmouth sail is still compulsory – the only working sail left. Poaching was a serious problem and many a Thames sailing barge coasting over the oyster beds with the tide under her keel and a load of bricks or wheat in her hold unobtrusively trailed an oyster dredge. Also, heavy-duty moorings were laid on the beds where watch boats spent the nights and weekends to keep guard. Massive lines of chains or 'creeps' were laid over the beds to catch dredges used by poachers.

For the small smacks of the Thames estuary, economic conditions were harsher than for the larger trawling smacks and herring boats of the North Sea. Many fishermen could not afford leather sea-boots and went bare-footed, binding their feet as best they could in winter. If the oyster fishing was bad in mid-winter they went stow-boating or dragged over the oyster-beds for 'five fingers' – a kind of starfish which preyed on oysters and could be sold as manure – using a dredge of worsted cloth which entangled the starfish in its nap.

Many Essex and Suffolk smacks ranged as far away as the Solway Firth and the Channel islands for oysters. Despite the Napoleonic Wars an oyster 'gold-rush' took place in 1797 when three hundred smacks fished the rich oyster beds discovered off Jersey. In his excellent book about the old-time sailors of the Essex and Suffolk rivers, *Northseamen*, John Leather has detailed many of their exploits – like the crew of one smack who were captured in the Channel by a French warship and spent seven years as prisoners before stealing a boat from Gravelines, rowing to Portsmouth and walking home to Brightlingsea. Many were fired on by French naval patrols for poaching and in 1828 a whole fleet of three hundred smacks stormed a French man-o'-war with hatchets and windlass staves, killing many of the crew, to free a smack that had been captured. To the dismay of local people, Essex and Suffolk dredging smacks cleaned out the oysters on the north coast of Norfolk, Jersey, and Largo Bay on the coast of Fife where Scots fishermen attacked them with stones.

When the railways joined the ports of the east coast rivers with London, numbers of large and handsome smacks were built by investors; one firm at Brightlingsea built thirty-six between 1857 and 1867. These were often called 'skillingers' because they worked off the Terschelling grounds near Holland, bringing back ten thousand oysters after a twelve-day trip. Until the last these smacksmen resisted labour-saving devices such as steam capstans, and all sails and dredges were man-handled. The heavy dredges were hand-winched up from twenty-five fathoms, muscle-cracking toil, and unpleasant, too, with the slimy three-inch warp embedded with razor-sharp shells it had picked up from the sea-bed. Out of the oyster season skippers found work where they could get it, carrying herrings from Stornoway to the Baltic, cattle from the Channel Islands to Weymouth, ponies from Norway to Newcastle and coals from Newcastle to London. Some boats ran from the west coast of Ireland to Liverpool with boxes of fresh salmon, a hard trip exposed to the full fury of the Atlantic, making the return trip in the incredible time of four days.

Yachting brought a limited prosperity to ports such as Rowhedge, Brightlingsea, Wivenhoe, Tollesbury and West Mersea in the latter part of the last century, and up to the Second World War. The winter stow-boaters and smacksmen became the summer crews of large yachts raced by the wealthy. There was also a tradition of racing smacks in fishing trim. The intensity, skill and fine judgement with which these working boats were pitted against each other by men accustomed to handling immense racing cutters in international events such as the America's Cup made thrilling racing. The tradition continues today in a small way with the annual barge matches and 'gaffer' races, but the smack races never really recovered after the First World War.

Bogs and Hoggies

The eastern end of Brighton beach, beyond the lines of bathing machines and the white sails of pleasure boats, was until about fifty years ago the centre of a small fishing industry that had flourished for centuries. Old boats sawn in half were propped against the sea-wall with bow or stern pointing upwards to provide storage huts for fishermen's gear. There were three distinct types of fishing craft drawn up on the beach, which was then flat and sandy, and most were used in turn through the different fishing seasons. One of them was the hog-boat, or hoggie, which was peculiar to Brighton.

Rather short and very beamy, with a sprit mainsail and sometimes a mizzen, the hoggie was a flat-bottomed boat that displaced eight to twelve tons and had leeboards in the Dutch style. Ballast of beach stones in bags was trimmed from one side of the boat to the other by the crew of three. Mostly, the sixty-odd hoggies working from Brighton beach during last century trawled local waters, sometimes working only three miles offshore or sailing to the rich area between Rye and Hastings known as the Diamond Ground where as many as three hundred sail of English and French trawlers could be seen working at the same time until the area became fished out in the 1850s.

Some hog-boats joined the larger three-masted luggers of sixteen to twenty tons displacement which had crews of up to seven men and specialised in drifting for mackerel. Early in spring the fishermen readied their mackerel nets, which were longer and had a larger mesh than the herring nets. When word came from Cornwall that mackerel were being caught they set sail, making Plymouth in thirty hours in a fair wind. Their catches were sold in Brighton, being landed by small

---

55. In a flurry of breaking water a lugger runs herself ashore at Hastings; in a few moments she will be hauled up the beach by capstan.
56. Herrings being auctioned on the beach at Hastings in front of a pair of tall tarred net houses.

56

57

HALLSANDS

boat or in calm water unloaded directly on to the sands, and as the mackerel came up the Channel with warmer weather their voyages grew shorter and shorter. At the end of August the season ended and the luggers were hauled up the beach by capstans. Then the small two-masted luggers of about six tons, and some hog-boats, were launched to fish for herrings until the end of the year.

Seaside visitors had to rise early to witness the fish auction on the beach, which began at six o'clock with the ringing of a bell by a man standing on the road above. Four to five hundred people, fishermen, buyers, sight-seers, milled round the salesmen as the fish were sold by Dutch auction. Most of the catch went by rail to London or was bought for the local hotels.

To launch a boat from the beach an anchor was carried out as far as possible at low tide, then the boat was run down on 'troughs' or skids, being checked when it went too fast by means of a shock anchor over the stern; as it floated the crew kedged it out through the shallows then recovered the anchor. When landing it rode into the beach on a big wave and was grabbed by other fishermen as the wave retreated. Then a bridle was quickly attached to the stem and it was drawn up by a capstan turned by a horse.

Most of the hog-boats were sold for firewood, and for the value of their copper nails, at the turn of the century but the last three Brighton luggers, which had been lying on the beach for years, were destroyed by the Admiralty during the Second World War so they could not offer cover for enemy soldiers if they invaded. The luggers were markedly similar to the beamy and bluff-bowed luggers of Hastings, which were supposedly adapted from the swift French privateers that harried English shipping until the end of the Napoleonic Wars.

At Hastings the three-masted rig persisted until the 1870s when the middle mast was dispensed with and drop keels, or centreboards, replaced leeboards. Hastings had been a fishing centre since the end of the Saxon period and was the most influential of the Cinque Ports in the Middle Ages; in the middle of the last century it was the biggest fishing centre where fishing boats were launched from an open beach. The eight-man luggers and smaller 'bogs', which were similar to the Brighton hog-boats, sailed to the North Foreland for herrings and to Cornwall for mackerel. Fishermen stored their nets in the tall, tarred 'net houses' which are still in use on the stade below the cliffs. At

57. Wicker crab pots being made at Hallsands, near Start Point, Devon; the village was washed away during a storm early this century because too much gravel had been dredged offshore for building Plymouth breakwater. Here Labrador dogs were trained to swim out with ropes to meet incoming fishing boats.

Hastings the beach was steeper and open boats ran the risk of being pooped and swamped by the waves; when this happened they were likely to be swung broadside-on to the waves and pounded to pieces. To lessen the danger they were built with beautifully elliptical and 'lute' (backward curving) sterns which also provided additional lift in a following sea. If a boat was pooped a man ran down the beach with a maul and as the wave receded he knocked in a plank as low as possible so the force of the succeeding waves was dissipated by letting the water run in and out.

Ramsgate had been an important fishing port since it was colonised by Brixham trawlermen in the 1820s and at the turn of the century it had a fleet of two hundred boats which fished local waters in the southern North Sea. Smaller boats trawled the waters round the Goodwin Sands for sole, plaice, skate, brill and turbot. Because of their proximity to France, Ramsgate fishermen were frequently in trouble with French boats which came right inshore in places such as Pegwell Bay to dig for bait when their own grounds were closed. Collisions were not infrequent and disputes were settled swiftly by an informal tribunal comprising a representative of the fishermen's mutual society and the commander of the French fisheries patrol gunboat who sailed in for the occasion. The larger French boats were not above towing chains to cut through English drift nets but they prized dogfish and often exchanged them for soles and turbots which fetched a high price in England. Ramsgate fishermen stayed in sail until many of their smacks were sunk during the First World War.

The trawler fleet at Dover dispersed when the War Office took over the port in the 1880s and built a large breakwater designed to hold the entire naval fleet. Nearby Folkestone had been a fishing port since ancient times. Daniel Defoe described how its fishing boats caught mackerel which were collected by Essex boats that went down to buy them and 'fly up to market with them, with such a cloud of canvas, and up so high, that one would wonder their small boats could bear it and should not over-set.' In 1900 the port had 50 sailing trawlers and a similar number of drifters that fished for mackerel in summer and herring in winter. Something of a record was set in September 1903 when a single drifter landed 20,000 mackerel and three others followed with nearly 50,000. At that time mackerel fetched five to six shillings a hundred but in such a glut the price soon dropped.

A mixed lot of smacks sprat-fished or dredged for oysters from ports such as the Hamble, and in every creek round the Solent there were small boats fishing with hand-lines and seine nets, their crews working in summer as paid yacht hands. Bournemouth did not have a single fishing boat at the turn of the century and, like Cowes, soles for its summer visitors came from Grimsby by rail. A few small boats trawled from Poole and some Devon ports such as Beer but their catches were

Tales of the deep – a long-shore fisherman and his pots.

small. Selsey was an important centre of crab, lobster and prawn fishing. The beach was dotted with tarred huts and piles of pots, but bait had to be obtained from Grimsby and cost each boat ten shillings a week. In fact, no other stretch of coast was as unproductive of fish as the 125 miles between Selsey Bill and Berry Head (Tor Bay); every little hamlet along the shore had a handful of small boats for setting pots and catching a few fish by hand-line, but few fish were caught in commercial quantities.

Mackerel Metropolis.

Fishing in the west country was precarious because the main catches were of migratory pilchards and mackerel which appeared off the coast in immense shoals, but also deteriorated more quickly than any other fish. The result was that until the railway crossed the Tamar in 1859 most fish could only be sold locally, and often there was such a glut that they were left decomposing in heaps along the foreshore. On the other hand the shoals may not turn up at all because of subtle changes in the temperature of the water. The characteristic of the west country fisherman's way of life until the middle of last century was long periods of enforced idleness interspersed with short periods of feverish activity.

It used to be said that if the chains of the Torpoint ferry at Plymouth were to break Cornwall would drift away because even if some of its boundary is river and stream, Cornwall is almost entirely surrounded by water. Until the middle of the last century Cornwall might as well have been an island, so remote was it from the rest of England. Its steep and twisting lanes, which today become choked with traffic at holiday time, were then difficult enough for horse and cart. Agriculture was insignificant, prosperity brought by the brief mining boom was limited, and until the tourist industry gathered steam after the First World War there was no real alternative livelihood other than fishing.

Devon was richer in agriculture and lay that important bit nearer to centres of population such as Bath, Bristol and London; it also had quite large residential populations of its own, such as Torquay which had become fashionable as a home for the families of naval officers in the blockade of France during the Napoleonic Wars.

On the northern coast of Devon and Cornwall fishing was confined to a handful of villages such as Ilfracombe, Port Isaac and St Ives where boats were afforded a little protection from the ocean surf by the natural lie of the land. Most fishing was done from the more sheltered southern coast where every river-mouth and every cove between ramparts of rock was put to use as a harbour. Deep and

Colleen Bawn
Mounts Bay mackerel driver 1897

narrow combes running down to the sea provided warmth and shelter
for fishermen's cottages which formed huddled villages that straggled
along the lines of twisting streams, their thatched roofs tied down
with chains against the blast and suction effect of winter gales. Houses
in what are regarded today as 'typical' fishing villages were built close
together for warmth and company, and also because fishermen needed
to live as near as possible to their boats. Streets were often merely
alleyways, wide enough to permit a man to carry a bundle of nets but
too narrow for a pack-horse. In Polperro men carried loads through the
village on poles slung over their shoulders.

The coast had an unsavoury reputation for deliberately wrecking
merchant ships for the sake of plunder. In the seventeenth century
land-owners objected to the building of lighthouses because their
opportunities for profit would be reduced. There are stories of lanterns
being tied to the horns of cows on a hillside to resemble the lights of a

village. It was also a smugglers' coast, and it was the need to hoodwink the revenue cutters that developed the fast, three-masted, half-decked luggers, about forty feet long, in the middle of the eighteenth century. These were the forerunners of the three-masted fishing luggers, of which there were some four hundred in Mounts Bay alone in 1808 – wonderfully seaworthy vessels, dry in a head sea and easy of motion because they carried no weight in their ends. The middle mast was used only in summer or when a group of vessels had a surplus of fish and one of them ran to market at Plymouth. If caught in a bad storm a raft was made of spars and nets and launched to windward to act as a sea anchor which also broke the force of the waves. In 1848 a small open lugger from Penzance sailed to Australia with five hands for the gold diggings; it carried mails from Cape Town to Melbourne, rode out a hurricane, and arrived in 117 days, faster than the average time taken by the regular packet.

The luggers were robustly built and some remained fishing a century or more. East of The Lizard where harbours were larger fishermen could afford to have boats with wide transom sterns that provided more deck space, but west of The Lizard the small, cramped harbours demanded double-ended boats which could nudge snugly up to each other like sardines in a tin and use less space. For cooking each boat had a hearth made of hollowed granite, or a box of clay and stones. Until the 1890s conch shells brought from the tropics by merchantmen served as foghorns. Sometimes they were filled with water to make a harsher sound; fishermen could judge their distance from cliffs by listening to the echo. Collision with fast steamers and square-riggers making for Falmouth or heading to or from the Channel was a constant hazard. For some unknown reason a fathom was counted as only five feet in some Cornish fishing ports and this led to such confusion that in 1868 an Act of Parliament had to define the fathom officially as six feet.

Boats did not have to be speedy because there was no fish train to catch. Mackerel were sold mainly to 'jowders' or 'jowsters' who, like the Newhaven fish-wives, tramped dozens of miles hawking fish to outlying farms and parishes. The fish was carried in a basket called a cowal which was supported by a broad plaited head-band. Drift-fishing was seasonable but it could be carried out nearly all year merely by changing the nets – mackerel from February to July, pilchard from July to December, herrings in December and January. Housewives cured their own supplies of fish, laying in about a thousand herrings and keeping them in large earthen jars or hanging them to dry in the sun. Pilchards were also home-cured but mackerel, obtained for eight to twenty a penny, were eaten fresh. Sprats were netted in the estuaries and crabs and lobsters caught in pots set off the rocky shore. Long-lines were employed from small boats fishing only a few miles out for gurnard, whiting and pollack. Two men fishing on a moonless night could fill a boat with nearly half a ton of conger eels, some seven feet long with a two-foot girth and turning the scales at more than seventy pounds. Conger pie was a popular west country dish and most cottagers made conger douce by slicing the fish lengthwise, sewing the pieces together to form a continuous length, and drying them on frames – a process described as 'most offensive to the noses of townspeople.' The sun-dried eel was then stored, and used grated to enrich soup.

The living style of west country fishermen at the end of last century was eloquently described by Stephen Reynolds: 'The kitchen is an extraordinary room, fit shrine for that household symbol, the big enamelled tin teapot. At the north-west corner is the door to the scullery and to the small walled-in garden which contains – in order of importance – flotsam and jetsam for firewood, old mats, spars and rudders, and some weedy, grub-eaten vegetables. At the top of the garden is a tumble-down cat-haunted linhay crammed to its leaky roof with fishing gear. No doubt it is the presence everywhere of boat and fishing gear which gives such a singular unity to the whole place.

'The kitchen is not a very light room. The courting chair stands in the north-east corner below shelves laden with fancy china and souvenirs – and tackle. The kitchener, which opens out into quite a comforting fireplace, is let into the east wall and close beside it is the provision cupboard so situated that the cockroaches, having ample food and warmth, shall wax fat and multiply. There is an abundance of chairs, all more or less broken. Over the pictures on the warm west wall (against which, on the other side, is the neighbour's kitchener) a line of clean underclothing is hung to air. The dresser is littered with fishing lines. Beneath the table and each of the larger chairs are boots and slippers in various stages of polish and decay. Every jug not in daily use, every pot and vase and half the many drawers, contain lines, copper nails, sail thimbles and needles, spare blocks and pulleys, rope ends and twine. Beneath the ceiling of the passage are laid masts, spars and sails. But most characteristic of the kitchen (the household teapot excepted) are the navy-blue garments and jerseys drying along the line and flung over chairs. An untidy room – yes. An undignified room – no. Kitchen, scullery, eating room, sitting room, reception room, storeroom, treasure room and at times a wash-house – it is an epitome of the household's activities and a reflection of the family's world-wide seafaring.'

The pattern of fishing in Cornwall changed completely after 1859 when Brunel completed the railway bridge over the Tamar at Saltash. A market train now left Penzance for Paddington at 2 p.m. on every week day. The fragmented cottage industry that was fishing in Corn-wall largely ended. With daily access to the great emporium, fishermen could now put their work on a proper economic footing. The harbour at Penzance was unsuitable because it was not accessible at all states of tide so catches were landed on Newlyn beach and taken two miles to the train by cart. Each lugger kept a large dinghy at Newlyn in charge of a boy aged between ten and fourteen who was known as the yawler. His job was exciting and difficult; this account by an old fisherman was quoted by Edgar March in *Sailing Drifters*: 'When the fleet was sighted in the offing all the yawlers sculled away to the best position for bearing down before the wind to "take" the lugger and the greatest rivalry resulted from each lad's efforts to be first alongside his special lugger, perhaps his father's boat. It was no easy job to take station to pick up a lugger smoking along, but one likely to teach the handling of an open boat in a seaway. With his boat pointing the same way as the big boat, the yawler waited with the coiled painter in his hand, ready for the critical moment when, with foam piling up at her bows, the lugger rushed by. As she did so the painter was thrown to those on board, and the lad crouched down in expectation of the jerk when the rope came taut and the boat sat up almost on end with water boiling up behind the stern.' As the lugger anchored the yawler rushed the skipper ashore with samples of the catch in a bucket to show the prospective buyers.

Drifters in Devon and Cornwall were known as 'drivers'. When mackerel driving, fishermen learned to perceive different shades of colour caused by plankton in the water. Yellowish 'cow-dung water' brought the best catches because it signalled concentrations of copepods brought out by lots of sunlight in February and March. Powerful swimmers, beautifully camouflaged by the dark ripple pattern on their backs for concealment in the sunlit upper waters, mackerel are of the same family as bonito, tunny and albacore which are the traditional Mediterranean fish. There was a conspicuous irridescent tint under the gills which indicated a mackerel's quality: old fishermen said green mackerel was poison, white mackerel was mackerel, but red mackerel was good mackerel. As they were sold individually and not by weight, fishermen picked out the biggest for their own use and their friends. This gave rise to the notion that you should not eat a mackerel unless it was given to you – because that was how you got the biggest and best fish. At the turn of the century a mackerel fetched anything from a farthing to a shilling at the boat's side (2,000 was a tolerably good catch) but the average price was nearer twopence.

---

58. *Cornish mackerel drivers had square sterns west of the Lizard, where the harbours were large, but pointed sterns to the east so that more boats could fit into the tiny coves and harbours.*

59. *Repairing the rudder post of a large trawler at low tide in Brixham Harbour; most smacks had to lie to large buoys outside the harbour because there was insufficient room for them.*

Like the Scottish herring boats, Cornish mackerel drivers became bigger and ventured further afield, often sailing to Whitby and Scarborough for the summer herring season but ignoring the East Anglia season and returning home in time for the November pilchards which may or may not turn up. A lot of boats spent the summer fishing from the south coast of Ireland and from the Isle of Man. When the mackerel fleet landed its catch at the Scilly Isles the fish were sent to France by the daily steamer and there were mad scrambles to get the fish aboard before sailing time. A large fleet of Breton fishing boats potted for crabs and lobsters near the Seven Stones reef. Their catch was sent to France every day by carrier and they frequently took refuge in St Mary's on the excuse of 'stress of weather' but the islanders did not mind because the French were said to spend more in a week than Scotsmen did in a year.

While the old-fashioned Cornish fishermen stayed loyal to the lug rig and drift-net fishing, Brixham in Devon was the birth-place of modern trawling. Although its fleet was the largest on the south coast the small harbour had none of the facilities that trawlermen enjoyed in the Humber. Only twenty-five vessels could get into the harbour at neap tides and there were no tugs so the big smacks moored to buoys and catches were ferried into the big quayside market by punt. Until about 1870 the fish was auctioned by women, and fish-wives played an important role in assisting their husbands. When the Parliamentary Commission investigating fishing sat in Brixham in 1882 the chairman asked a fishing skipper, 'I am satisfied you have no disputes about wages, but I cannot make out how it is done.' The skipper replied, 'We leave all that to the women.'

When the railway reached Brixham in 1868 there was a feverish boom. People sold their houses to invest in new smacks and ship-building in Devon, which had been at a low ebb because of the failure of the fruit trade, suddenly reached a new though short-lived peak of prosperity. The sandy bottom of Tor Bay was ideal for beam trawls and famous for its soles so the smaller smacks called mules and mumble bees did not have to go far to fish. The larger smacks fished the Bristol Channel, Dublin Bay and Liverpool Bay, landing their catches at the most convenient port, so they were away from home for eight months of the year. A tradition of the port was the great regatta, held on the first Friday and Saturday in August when the smacks had returned.

In January 1866 a gale sank more than sixty vessels in Brixham and Tor Bay at a cost of more than a hundred lives. Huge seas carried away the lighthouse on the end of the pier, which was replaced by an enormous fire lit by women of the town and kept going with piles of old nets and barrels of tar. Wrecks were driven right into the main street. The 1200-foot breakwater which guards the harbour today was built with financial help from fishermen who subscribed threepence in the pound of their income, and it was not completed until the First World War.

Trawlers also worked from Plymouth but were constantly 'at war' with the Royal Navy whose firing ranges overlapped the best trawling grounds. Several smacks were the victims of near misses, and in 1891 *HMS Plucky* succeeded in sinking two smacks during firing practice; the officer in charge was found guilty of negligence for incorrectly estimating distance affected by peculiar conditions of the atmosphere. The Barbican was a perfect all-weather port and was used by fishing boats from other harbours to land fish. The building of the huge breakwater across Plymouth Sound indirectly caused the demise of the small beach-fishing villages of Hallsands and Beesands, near Start Point. Dredging for sand and gravel to build the breakwater so undermined the shore that during a fierce south-easterly storm and spring tide in 1917 both villages were washed away. Fishermen here had a unique method of landing their boats through the surf. They were met off the breakers by swimming labrador dogs which carried lines ashore to helpers who pulled the boats in.

Strangely enough it was the building of a harbour at the 'supreme mackerel metropolis' – Newlyn – which started the decline of Cornish mackerel driving. The harbour allowed big steam trawlers to land fish, and every spring and summer more than a hundred smacks and steamers trawled day and night over the best mackerel grounds. Skipper Tom Jenkins of the drifter *Golden Fleece* told an inquiry at Penzance Town Hall how trawlers affected his fishing on just one night when he was two miles south-east of Wolf Rock lighthouse. At 10 p.m. he saw a trawler on the port beam and tried to haul his nets clear without success. 'Our nets got foul of her trawl ropes so we hauled up alongside her and round her bow and got clear. Then we saw twenty or thirty smacks outside us and after a bit found another trawler foul, hauling his trawl. We sung out to him to stop as he was destroying us but he didn't hear at first. The tide was running so strong we couldn't get round his bow so we had to unbend some of the nets and cut the warp. After that three other trawlers which we didn't catch cut their way through our nets with axe or knife deliberately and made our warps fast again afterwards. We lost a night's fishing and several nets worth six or eight pounds.'

Like the North Sea trawlermen who took the gospel preaching of the missionaries to heart, west country fishermen were greatly influenced by John Wesley's revivalism and for many years refused to fish on a Sunday. The trawlers that came down from the east coast, however, worked seven days. Whether the Cornishmen's anger was due to religious fervour, resentment against trawlers ruining their nets, or the fact that Sunday fishing by strangers spoiled their Monday market, is not known. But in 1896 their anger finally boiled over.

On a Sunday evening in May a few Lowestoft drifters came alongside Newlyn pier ready to sell their catch next morning. A crowd of fishermen, mostly from nearby Mousehole, many of them drunk and goaded by fish-wives, pushed past a small and anxious guard of coastguardsmen, surged aboard the drifters and threw one hundred thousand good mackerel into the harbour. Then they seized three other boats lying outside the harbour and did the same, stretching a heavy chain across the entrance.

The event was likened to the Boston tea party. The uproar lasted two days. Extra police were drafted in by rail. Soldiers were called out. The Navy sent a gunboat and a destroyer, which anchored off the harbour. Finally, after a lot of heads had been cracked, the chain was withdrawn and the east coast fishermen agreed to refrain from fishing between Saturday and Sunday night. Only three Cornish fishermen were charged, and were sentenced to be bound over. The Board of Trade assessed the damage at £619 15s. 3d. and peace was restored. But the drifters, their repeated requests for gunboat protection ignored, lost their war against trawlers and the traditional Mounts Bay luggers slowly faded away.

Meat, Money and Light.

Like the shadow of a cloud moving slowly across the water the shoal of pilchards was spotted by a lookout posted in a cliff-top hut. 'Hevva! Hevva!' he cried, alerting the town which immediately mustered. With the vigour and excitement of a lifeboat-launch the heavy tarred boat was directed to the shoal by the 'huer' on the cliffs using a kind of semaphore with the aid of a furze bush in each hand or a 'bat' made of white calico stretched over a bent hoop.

Pilchard-seining was a colourful and extraordinary sight unique to certain parts of Cornwall and south-west Devon. It was quite distinct from the catching of pilchards, with drift-nets by luggers. The aim was to surround an entire shoal of thousands of fish and haul it bodily into shallow water where the net was moored and the fish remained alive until required for curing. The pilchard is better known as the sardine: it is a sub-tropical fish, similar to the herring and the west country is at the northern limit of its territory.

Seining had been carried on in much the same way at least since the Middle Ages. Elizabeth I, James I and Charles II all enacted laws to protect the seines, mainly from drift nets which could prevent the huge shoals from approaching inshore. By late Elizabethan times pilchard exports played a significant part in the country's international trade and precautions had to be taken to ensure that enemies, notably Spain, were not supplied. When smoked for export to Spain pilchards were known as fair maids, a corruption of the Spanish word *fumados*. Later the traditional market became Italy which took huge quantities of salted pilchards in casks for consumption during Lent. In the 1870s some curers packed decayed fish with fresh and it was years before the reputation of Cornish pilchards recovered.

In Richard Carew's Survey of Cornwall (1602) 'sayning' was described and the method remained unchanged for another three hundred years: 'The boats lie hovering upon the coast and are directed in their work by a balker, or huer, who standeth on the cliff side and from thence best discerneth the quantity and course of the pilchard: according whereunto he cundeth (as they call it) the master of each boat (who hath his eye still fixed upon him) by crying with a loud voice, whistling through his fingers, and wheazing certain diversified and significant signs with a bush which he holdeth in his hand.'

When a shoal was sighted it was an anxious time until the catch was landed, for a second chance might not come for a whole season. First the shoal was surrounded by the seine, a fine-mesh net at least 160 fathoms (960 feet) long and 8 fathoms (48 feet) deep. The operation was directed by the master seiner in a small boat called the lurker. While the net was drawn round the shoal a third boat, the 'volyer' or follower, set a 'stop' or 'thwart' net across the gap in the circle. Boys in the lurker beat the water with oars to prevent the fish escaping before the ring was closed. Meanwhile a huge three-inch warp, nearly a mile long, was landed on the shore nearby and about twenty-five men called blowsers stood by the nearest available capstan on the foreshore to warp the entire seine with its enclosed shoal out of the tideway and into shallow water where it was anchored with grapnels. At low water the seine boat rowed inside the net and with a small net called a tuck scooped the pilchards up and loaded them into small ferry boats called dippers which rushed them ashore.

When half a dozen seines had been shot in succession the hustle and bustle ashore was incredible. With escorts of small boys armed with sticks paid to prevent other small boys from snatching pilchards as

A large catch of pilchards caught in drift-nets and being unloaded at Newlyn, Cornwall.

they went by, the fish were carried through the streets in gurries – open boxes carried between two shafts like a sedan chair – to the curing cellars or pilchard palaces. These were large rectangular buildings with thick stone walls and open courtyards. The floors were covered in pebbles set in concrete and sloped inwards to a gutter which led to a pit in which a large cask was set. At amazing speed women stacked the pilchards between layers of salt into square heaps five feet high where they were left for a month to let the oil, salt and blood drain away. The salt was kept for use again, and the oil was skimmed out of the barrel. Later, the pilchards were washed then carefully packed in casks called hogsheads, which were wider at top and bottom than ordinary barrels. Boulders were put on the lid to squeeze any remaining oil out through the bung-hole and the barrel was topped up from time to time; when ready for export it held about three thousand pilchards.

The seine could take a week to empty and was guarded overnight. One was moored long enough for a lugger to race to Britanny for salt. But there was always a risk of bad weather leaving the fish rotting in heaps along the foreshore. In 1871 a seine containing twenty-four million fish was lost in this way. Then farmers carted them away for a few pence a load. The benefit of pilchards used to manure land lasted as long as seven years. But while the drift-net fishermen lived at subsistence level the seine owners could make a small fortune – although they risked a lot because their men were paid by the season and sometimes shoals did not turn up for years at a stretch. In October 1862 a seine took 3,500 hogsheads which sold for £3 15s. each: after expenses for packers, labourers and fishermen this left a profit of £10,153. Fishermen took a small share of the catch for their own consumption, and dried or pickled them in their cottages. Pilchard oil was used for painting iron, treating leather seaboots which became stiff in salt water, and as a fuel for rush lanterns: Cornishmen said pilchards provided 'meat, money and light – all in one night.'

The excitement of a pilchard shoal was recorded by Daniel Defoe early in the eighteenth century who was being rowed to inspect Dartmouth Castle when he saw some fish skipping and playing on the water. 'Immediately one of the rowers starts up in the boat, and throwing his arms abroad as if he had been bewitched cried out as loud as he could bawl, "A school, a school!" The word was taken to the shore as hastily as it would have been on land if he had cried fire; by the time we reached the quays the town was all in a kind of uproar.' A shoal of pilchards had swum into the harbour and Defoe's merchant friend observed that with a little more warning they could have taken two hundred tons of them. As it was, only a few thousand were taken by one or two small boats, and when Defoe sent his servant to buy some for dinner he was astonished to get as many as he could eat, including the cost of cooking, for only a farthing.

"May the streets run blood and the cellars run train oil!" This was the fervent prayer of Cornish pilchard fishermen, here surrounding a shoal with an enormous seine net. Once the shoal was encircled the whole thing was winched into shallow water and anchored, then the fish removed as needed with small dip nets.

Traditional seining centres like Cawsand, Fowey, Polperro, Mevagissey, Newlyn and Newquay stopped this method of fishing in the middle of the last century, partly because drift-net fishing interfered with the shoals and partly because the railways allowed higher prices to be obtained for other fish with less risk. The once substantial industry now became centred on just one place – St. Ives. This most picturesque of Cornish fishing ports had always been the 'capital' of pilchard seining and its catches were often greater than those of all other ports combined. Its coastline was divided into six areas called stems which were fixed by statute in 1848 and marked by white poles on the shore – Porthminster, The Poll, Leigh, Carrack Leggoe, Carrack Gladden and Pedn Olver. There were so many seines that boats had to take turns working in the different stems and some years a seine-owner was lucky to draw more than one day's fishing every seven weeks during the six-month season.

The seines had names, like boats – *Lion, Neptune, Speedy, Anthony and Rose, Gannet, Plover, Speculation, Good Heart,* and so on. At its peak in 1869 St Ives had 286 seines owned by 36 individuals or companies. The largest owners were T. Bolitho & Sons with 33. The black-tarred boats had distinguishing bow marks such as a white ball enclosed in a white ring, or three horizontal white bars. Boats were not allowed to chase a shoal into another stem but could shoot a part of their net over the boundary. Once the shoal was enclosed it could be worked through and moored in any other stem.

The number of seines declined sharply in the 1870s as the shoals became infrequent. Blame was laid on drift-net fishermen who could be prosecuted for shooting nets too close to the shore, but seine-owners were dependent on drift-fishermen's wives to cure the pilchards and as it would have been disastrous for a drift-fisherman to have his nets confiscated by a court it was not in their interests to press for prosecution. It was also said that pilchards formed shoals in shallow water only to escape from hake which were caught in great numbers by steam trawlers in western waters after the 1890s, and the decline of these predators enabled the pilchards to remain safely in deep water.

By the turn of the century there were only two seines working but the number increased briefly to forty-five in 1904 after six million pilchards were caught in a single day. The last catch was made in August 1908 but the seiners continued to wait for the return of the shoals until 1922 before they gave up. It was the belief of one of the old huers that one shoal which had come within two miles of the shore in 1908 had been frightened by a train whistle. This could have been true because previously shoals had been affected by the sound of guns at target practice.

Nickeys and Nobbies.

More by default than design the most industrious herring fishers of the Irish Sea were Manx fishermen. For the most part the independent Isle of Man government avoided the swingeing salt taxes which for centuries crippled the efforts of fishermen on the west coast of England and Scotland. As early as 1610 the enlightened Manx government enforced a January to July close season for herrings within nine miles of the shore, and prohibited the shooting of nets before sunset – or, 'while your hand being extended to the uttermost distance from your body, you can still distinguish the black of your nail.' The law lapsed in the 1820s when the island's waters began to be fished by increasing numbers of English boats, mainly from Cornwall.

At this time the Manx fleet comprised about 260 boats crewed by 2,500 men who also crofted and spent much time gathering seaweed for manure. The boats at first had square sails and were called dagons, but later they changed to 'dandy' or yawl rig. However the lug rig of boats visiting from Cornwall was more efficient, though more difficult to handle, and Manx fishermen took the unfashionable step of converting back to lug rig. Their luggers were called nickeys because Nicholas was a common christian name among the Cornish fishermen.

Until the middle of the last century fishing in the Isle of Man was a curious mixture of drink, superstition and religion. In April boats were launched from the banks where they had lain upside down all winter. The skipper hired his crew at the 'boat's supper' which was held about Christmas time with lots of pies and puddings, rum and jough (cheap beer). The skipper passed a shilling to the man he considered the best and named his conditions. This man passed it on to the next, and so on. The last man put it in a quart pot, shook it, and up-ended the pot over the shilling: if the coin showed 'heads' it was a good sign. When fishing started in July the crew chose a pub where they started the 'shot' – a drink on credit. Any of the crew and their wives and sweethearts was able to have their pint or a glass of rum at any time and the account was settled weekly or at the end of the season.

When clear of the harbour all hands took off their hats and prayed, then rum was served in a horn measure that had been handed down from father to son for generations.

As they neared port with their catches they were met by 'bimming' yawls, carrying buyers. The buyers boarded the fishing boats and after striking a bargain handed over a shilling deposit called the earlys, or earnest, and passed round a bottle of rum'. The fish were run fresh to the English markets in smacks lying ready in the habour, or were barrelled and exported to the slave plantations of the West Indies. No matter how poor they were the Manx fishermen never worked on a Saturday evening or a Sunday. On Saturday the boats made for harbour and each man separated his nets which he carried on his back to the nearest grass and laid out to dry. Any odd shillings left over after dividing their money in the pub was called 'God's portion' and good luck depended on giving it to the poor.

These Manx customs were gradually lost as the island assimilated increased numbers of drifters visiting from the east coast of Scotland, East Anglia, and Cornwall. In the 1860s there were six hundred boats sailing from Manx ports and by 1880 the number was nearer a thousand. The island's own boats now were changed back to fore-and-aft rig and called nobbies, but the true 'nobby' was a smaller vessel, about thirty feet long, developed for trawling shrimps in Morecambe Bay. There, shallow muddy estuaries and extensive sand-flats exposed at low tide also provided work for scores of cocklers. At the turn of the century processions of cocklers could be seen following the tide as it receded, trudging out in long lines for as much as ten miles from isolated villages such as Flookburgh and Allithwaite on the northern shores of Morecambe Bay. Children were attached to their mothers in case they stepped in a hole; only the men had waterproof boots but the women were allowed to ride in the carts when crossing the deeper channels. The cockles were brought to the surface by an apparatus called a jimbo, like a small bench seat which was inverted and rocked back and forth on the sand; it was thought to have originated from rocking cradles put down on the sand by cockle-gatherers.

Fishing at Fleetwood was started between 1835 and 1840 when both the pilot boat and the local revenue cutter started carrying beam trawls with which the crews augmented their earnings. Then a railway company bought a few old trawlers from the east coast and by 1875 there were seventy smacks at Fleetwood and forty-one at Liverpool supplying the industrial towns and cities nearby. Fleetwood was close to supplies of good coal and its importance increased with the rise of steam. In 1892 several owners including Robert Hewett transferred from the east coast to obtain cheaper coal and just before the First World War the port had nearly 100 steam trawlers fishing as far afield as Rockall and the coast of Morocco.

There was a herring fishery at Aberystwyth in the Middle Ages and some Pembrokeshire fishermen sailed to Newfoundland for cod, but Welshmen seemed less inclined to go to sea. The 1913 government fisheries report commented, 'the sea instinct seems to be comparatively much less developed along the Welsh seaboard; long stretches appear to remain more or less unfished except by vessels from other coasts.' And at the turn of the century the fine harbour at Holyhead had had practically no fishing boats. Welshmen restricted their fishing activities to the foreshore, setting stake nets and weirs on the tidal sands. Many were set near the low-water mark three miles from shore, and in Swansea Bay fishermen raced the tide back with their equipment and their catch using a kind of narrow sledge, like a big ski eighteen inches wide and six or seven feet long. On this was a frame, like a low table. The fisherman lay on top of the frame and pushed with his feet, and it was said that on ground so soft that a man would sink he could travel as fast as a horse could canter.

Fish were landed for sale at Welsh ports such as Tenby, but almost exclusively by trawlers from Brixham. Local people were poor and did not muster more than a few open boats. In 1863 some fishermen worked inshore from Caernarfon but they came ashore when work became available at the slate quarries inland; trawling in Caernarfon Bay and Cardigan Bay was in the hands of trawlers sailing from Hoylake and Liverpool.

What Wales did have was great quantities of cockles and mussels in its beautiful estuaries, and one of the finest harbours of Britain situated on the very doorstep of the open Atlantic with easy access to good trawling and mackerel grounds – Milford Haven. In 1888 the Great Western Railway opened a fish dock one thousand feet long. At first there were not even any fish buyers at the port and when the first consignment was sent direct to Billingsgate via Paddington it caused a sensation. Here was another example of what a railway could do for a port that enjoyed considerable natural advantages. The 4 p.m. train reached London at 2.30 a.m., in plenty of time for the early market. In two years 10,000 tons of fish was landed; by 1902 it was handling 24,000 tons a year. The port was also well placed for supplies of best steam coal and in the 1890s it was the home port for about seventy steam trawlers while a fleet of about two hundred smacks from Brixham and Lowestoft trawled the Bristol Channel for soles between March and October.

60. *Shallow muddy estuaries and extensive sand-flats exposed at low tide provided work for scores of cocklers who followed the tide out.*
61. *Lowestoft herring lugger* Branch *about to sail on her maiden voyage.*

The Best Sauce is Appetite.

Yarmouth bloaters, Newcastle kippers, Finnan haddies, Auchmithie pinwiddies – when travelling time in Britain was measured in days rather than hours it was the smokehouse that filled the gap between the dreadful dullness of rough-salted and wind-dried fish, and the doubtful privilege of fresh fish (unless you could buy it on the quayside). Only the fishermen themselves could enjoy the finest recipe of all for cooking herring, the so-called Dutch way – transfer the fish directly from the sea into a kettle filled with boiling sea water. The traditional fishermen's sauce was a handful of mustard shaken up in half a bottle of vinegar with the thumb over the top.

At about the time the Dutch discovered the art of pickling herrings in brine, a Great Yarmouth fisherman discovered the merits of smoking herring for a long time. The 'red' herring, chewy and tasting strongly of wood-smoke, was Britain's principal fisheries export for more than three hundred years and was also consumed at home. Known as the bacon herring, Glasgow magistrate, or militiaman (conversely, Glasgow magistrates and militiamen were often known as red herrings) a properly cured red herring's skin should peel off easily. Then the head was removed and the herring was cut across into four pieces, dusted with pepper, and eaten with a slice of bread. For the fishermen the advantage of selling to red-herring curers was that little salt was required and quality was more or less immaterial so the boats could stay at sea for several days and return with a good haul.

The British palate always favoured cold-smoked fish which were not even partly cooked, such as kippers, bloaters and Finnan haddies, while continental people favoured hot-smoked fish such as buckling and red herring which were cooked in the smoke.

The kipper was deliberately invented by John Woodyer at Seahouses in Northumberland; kippered salmon was the forerunner of smoked salmon, and after several years of experiment the method was adapted successfully for herrings by Woodyer. The first kippers sent to London in 1846 were instantly popular, and soon became as traditional a part of the British diet as roast beef.

While the red herring was steeped in salt for two days and smoked for two or three weeks with intervals for the oil to drip out, the kipper was split and gutted, soaked in brine for about half an hour, then hung on hooks from long rods called tenters in the smoke of a fire of oak chips and sawdust for six to eighteen hours. The result was a darkish brown herring with a sweet and nutty flavour. The cut surface and skin of a good kipper should be quite oily, the flesh firm and springy but fairly rigid. Old-time curers were very particular about their products and when vegetable dyes began to be used during the First World War to produce the dark look of oak smoke and prevent the fish from losing weight during curing the traditional curers could no longer compete and by the 1930s all kippers were dyed and known contemptuously in the trade as painted ladies.

The silvery yellow Yarmouth bloater was roused in salt overnight then smoked over fires in which heat was more important than smoke so it was lightly cooked yet had a slight flavour of wood-smoke. The bloater had many admirers: 'Note, if you please, that the bloater is the *élite* of its class – the largest, handsomest, most fleshy and corpulent, in all respects the aristocrat of the herring tribe. Look at him as he lies in the dish tempting your morning appetite, mark his rich colouring rivalling that of an old master – the deep olive tones of the back, the metallic brilliancy of the sides, softening into a creamy hue in his under parts. Then what a fragrance he exhales as you lay him open and hasten, at the instigation of your salivary glands, to pay him the posthumous honour to which he is richly entitled.'[1]

The type of wood used in smoking was important. Shavings and sawdust of hardwood such as oak were preferable to softwoods which gave a good colour but were liable to impart a resinous flavour to the fish. Teak and pitch-pine produced an acid flavour, mahogany, red-wood and Oregon pine gave a good colour and a passable flavour. Good quality smoked fish was free of smuts and dirt and had a bright and glossy cut surface, but the art of smoking was getting the right combination of heat and smoke – too much heat cooked the herring and turned them into buckling, which were popular on the Continent eaten cold.

The Scots made a speciality of smoking white-fish such as haddock. Early in the last century these were thrown overboard by English fishing boats as worthless, but in Scotland they were a principal food-

---

[1] *Leisure Hour Magazine* (1868).

fish. Fish-wives smoked them in barrels half set into the ground, using the soft smoke of a low peat fire. The fish was slit down to the anus then the belly flaps were bent outwards and another cut was made alongside the backbone to let the smoke penetrate. In the 1820s women in the fishing hamlets along twelve miles of coast between Aberdeen and Stonehaven – where one village was called Findon – packed small quantities of these delicately cured fish into baskets and gave them to the guard of the 'Defence' coach as it passed on the way from Aberdeen to Edinburgh, and the guard passed them on to his brother who dealt in provisions. The fame of the smoked fish rapidly spread until 'Finnan haddies' were distributed to all parts of the country, first by steamboat and later by rail.

Smokehouses for curing haddocks were built at the big fishing centres such as Aberdeen, Grimsby and Hull, and haddocks were no longer regarded as 'offal'. But the commercial establishments used wood smoke instead of peat and purists complained that a Finnan haddie smoked over wood was no better than orange marmalade made with turnips. Meanwhile the genuine cure fell into disrepute during the 1870s because the fish-wives used bad peat that provided too much heat and not enough smoke, so the fish were partly cooked and difficult to handle; they also started using resinous sawdust that put colour quickly on the fish but produced a bitter taste.

With the coming of railways to Scottish fishing ports in the 1860s large numbers of smokehouses were built. One of the most popular methods was the Eyemouth cure which was used on most of the east coast from Berwick to Inverness, and near Glasgow. Haddocks were smoked a ton at a time for six to nine hours over smoke from soft peat and white-wood sawdust which gave the fish an even golden colour and a glossy look, like varnish. After smoking the fish was carefully wiped of 'coom' or ashes with a rag soaked in warm sea water and matched in bunches of three, tied with a dried rush, and packed in baskets. The process was intricate but the fish lasted long enough to be transported almost anywhere by rail. Their flavour was exquisite.

The Auchmithie cure produced smoked haddock that were also cooked and were known as pinwiddies. Their colour was a dingy brown or black but the smoke filtered through the skin and left the flesh deliciously sweet and perfectly white; most were cured by fish-wives using smokehouses outside their cottages. Bervies were haddocks cured

---

62. *Brine-soaked herring are threaded on "tenters" which are hung on hooks high up in the smoke-house and are cured and partly cooked by the smoke to become kippers.*
63. *Sousing herring in brine preparatory to curing it in a Great Yarmouth smoke-house.*

in the Findon area after the Finnan haddie became so successful. Fish-wives used stickly peat and 'yellow sods' (recently formed peat still covered with bog moss) which were added to the fire as the cure was completing so the flames blazed up and cooked the fish. The Moray Firth or Buckie cure was used when the market required it by curers making Eyemouths. The fish were left in the kiln much longer and came out well dried and a beautiful golden yellow, with the result that they stayed sweet much longer. The disadvantage was that the fish had to be rearranged in the kiln while the fires were still burning, to ensure an even cure according to their size, and this was disagreeable work in intense heat and thick smoke.

In Scotland the traditional method of grilling fresh herrings was to cut off the heads, tails and fins, split them open, season the inside with salt and pepper, then place two herrings flat together, backs outermost, and rub them in toasted oatmeal then fry them – often with plenty of sliced onions. According to the French *Almanack des Gourmands*, written early in the nineteenth century, herring cooked on a gridiron with melted butter and sharpened with mustard was an ideal method – 'But it is essential to observe that the fish requires only to see the fire to be cooked. It is truly disobliging it, and, worse than that, drying it up, if one leaves it long enough to repeat a Paternoster. An Ave Maria will suffice. You may eat fresh herrings also *en matelotte* but only as hors d'oeuvre, for the herring has not the pretensions to be raised to the dignity of an *entrée*.' This was the recipe for herring *en matelotte*: Gut six to eight herrings, take off the heads and tails, and cut each one lengthways in two parts. Sprinkle with salt and pepper. Place in a stewpan a little fresh butter stiffened with flour, a little parsley, and some shallots minced fine. Place the herring sides on this dressing, sprinkle with two or three glasses of red wine, and cook over a hot fire. Have ready a dozen small onions and a score of little mushrooms, cut small. Fry some breadcrumbs to garnish the dish on which the *matelottes* are served up.

In Germany, lightly pickled herrings in salad were considered a sovereign remedy for 'that throbbing, tingling, twingeing sensation of the brain and nerves, which, after a night of heavy potations at the drinking houses, is known to muddled students as the *katzen-jammer* or lamentation of cats; taken freely it *bringt der burschen auf der damm* – picks you out of the ditch and sets you up on the dyke again.

V

NET PROFIT

Fishing has become sad.
    (*Alongshore*, Stephen Reynolds, 1910)

Sign of Zodiac.

Steam power came slowly to fishing but when it did come it took over with astonishing speed. In less than a decade the face of the industry was totally changed. Steam spelled the end of individual ownership, concentrated the industry in a small number of ports, greatly reduced the number of people earning their livelihood from the sea, and soon had a devastating effect on the traditional fishing grounds.

Fishermen had been familiar with steam for many years but seemed blind to its obvious advantages. The paddle-wheeler *Savannah* had steamed across the Atlantic in 1819, sail was well into its decline in the merchant fleet, and smacks had been towed in and out of fishing ports by steam tugs since the 1850s. Engines tended to be unreliable and gobbled coal, and it might have been difficult to make them pay, but even when the first steam winch was seen at Grimsby in 1877 fishermen dismissed it as a novelty, although it was not long before practically every smack was fitted with an 'iron man'.

During the hot, calm summer of 1860 the smacks *Henry Fenwick* and *Fearnot* were towed about the inshore grounds off Sunderland by the paddle tug *Heatherbell* which joined them at noon every day five or six miles out, towed them with their beam trawls down until midnight, then steamed their catches ashore for the early morning market. They were so successful that other owners tried to cash in on the idea but the tugs were unsuited to the open sea and the project lapsed when the autumn winds arrived.

A significant development came in 1877 when many steam tugs were laid up in the north-east rivers because railways had taken over the huge coastal trade in coal, and it occurred to Mr William Purdy, part-owner of the tug *Messenger*, that he could just as easily tow the trawl himself and dispense with the smack. In his detailed book *Sailing Trawlers*, Edgar March tells how smack skippers roundly denounced the notion as 'absurdly new-fangled and hopelessly impracticable' and assured Purdy he would lose his money.

'The tug was ancient, built in Jarrow in 1843 with engines of only

Consolation, *the first steam-drifter, so named because rigging was provided in case of engine failure; built c. 1890.*

twenty-five horsepower, and to catcalls and jeering it sailed on November 1 1877 with a beam trawl put together from bits and pieces found around the dock, and trawl nets obtained from Grimsby for ten guineas. The first trip made £7 10s. and picked up a lucky £5 for towing a vessel in. The second trip saw tug owners falling over themselves to follow suit until not one idle tug swung to moorings. The following year fifty-three paddle tugs trawled the inshore grounds between the Tyne and the Humber, and Purdy was publicly presented with a gold watch by members of the fish trade.'

Still the smacksmen were not unduly worried: the steamers were unreliable, had only a short range, and were not the kind of vessel in which you would choose to encounter a North Sea gale.

Soon they had something more serious to ponder. In December 1881 the Great Grimsby Steam Trawling Company launched the first specially designed steam trawler which was based at Grimsby. It was called *Zodiac*, and a second vessel, *Aries*, followed immediately and sailed from Hull. If any man of sail had been able to see the changes which the arrival of these two vessels foretold he would have been incredulous. Ninety-eight feet long – considerably larger than any smack – and capable of steaming at nine knots with a consumption of four tons of coal a day, these screw-driven wooden steamers soon proved their value by catching four times as much fish as any of the sailing trawlers.

The result was an immediate and almost total conversion from sail to steam, and the increase in catches was spectacular. Grimsby sent away 47,000 tons of fish by rail in 1880, 71,000 tons in 1890, 134,000 tons in 1900, and 180,000 tons in 1910, when before Easter five miles of fish wagons were despatched in a single day. Grimsby fish market, with its covered roof more than a mile long beneath which up to 100 trawlers could discharge their fish, was the largest in the world. Big fish such as cod, halibut and skate were laid out in rows on the flagstones, while line-caught cod and halibut, brought home alive in steamers' wells, flopped and floundered about on the slime.

The coming of steam saw many auction notices like this, announcing the impending sale of smacks which could no longer compete.

TO BE

SOLD BY PUBLIC AUCTION

BY

MR. W. J. S. HOOD,

On board the Vessels in the Extension Fish Dock, Great Grimsby,

On TUESDAY, the 11th day of DECEMBER, 1888,

AT 12 O'CLOCK AT NOON PUNCTUALLY.

All that Dandy Fishing

VESSEL

CALLED THE

"FRIEND'S GOOD WILL,"

OF THE PORT OF GRIMSBY.

Built at Rye in the year 1881; Length, $73\frac{6}{10}$ feet; Breadth, $20\frac{5}{10}$ feet; Depth $10\frac{5}{10}$ feet; Registered Tonnage, $71\frac{86}{100}$ tons.

ALSO ALL THAT

DANDY FISHING VESSEL

CALLED THE

'MARSHALL'

OF THE PORT OF GRIMSBY.

Built at Burton Stather in the year 1874; Length, $72\frac{3}{10}$ feet; Breadth, $19\frac{8}{10}$ feet; Depth, $10\frac{3}{10}$ feet; Registered Tonnage, $71\frac{48}{100}$ tons.

A List of the Stores, which will be sold with each Vessel, will be produced at the time of Sale. *The whole or part of the Purchase Money can be obtained on approved Security.*

For further Information apply to the Auctioneer, or to

STEPHENSON & MOUNTAIN

SOLICITORS, GREAT GRIMSBY.

W. J. EDEN, PRINTER, VICTORIA-STREET, GRIMSBY.

247

In 1909 there were 1,336 steam trawlers in England and Wales – 514 at Grimsby, 446 at Hull, 76 at North Shields, 31 at Boston, 68 at Milford Haven, about 35 at Fleetwood, and small numbers at ports such as Swansea and Plymouth which did not long survive the competition; Scotland had 278, mostly at Aberdeen. The Humber ports combined had more steam trawlers than all the other countries with North Sea borders. Half of Britain's total catch was landed at Hull and Grimsby, and the share of more than a hundred minor ports dropped to less than two per cent.

Formerly the dispersal of fishing ports around the coast had been an advantage because boats powered by sail and oar had comparatively little stamina and they needed to be as close as possible to the fishing grounds. For steam a greater range and capacity of port facilities was required – deeper harbours, bigger and longer quays, bigger dry-docks, more workshop facilities because it was no longer possible to assemble fishing gear at sea, ice houses, coaling wharves. By 1910 the steam fleets were consuming 700,000 tons of coal and 200,000 tons of ice every year. Although trawlers were registered at other ports they tended always to land their catches at one of the six main centres, Hull, Grimsby, Lowestoft, Milford Haven, Fleetwood or Aberdeen.

Like Cornish fishermen the Scots had traditionally taken a dim view of trawling because they thought it damaged the grounds; despite a slow start there were 217 'scratchers' or steam trawlers and fifty-three steam-powered long-line vessels sailing from Aberdeen by 1910 and the port handled well over half all the white fish landed in Scotland. But fishing in other ports declined and many crofters laid up their small boats and shipped as crews in the steamers, usually men from one village going in the same ship.

Suddenly, the days were gone when a good fishing smack could be built for about £1,500 and an energetic man could become an owner with relative ease. Steam trawlers were four times the cost of a sail smack, too expensive for the ordinary man. In 1889 Grimsby had 292 owners with one smack each; 12 years later its steam fleet was owned by 69 different companies and only two or three were owned by single individuals. As the smacks were auctioned off for as little as £200, mostly to Dutch, Irish and Faroese buyers, it was not unusual to find three or four former skipper-owners shipping in one steam trawler.

The famous Short Blue fleet of Gorleston took over 90 smacks of the Grimsby Ice Company in 1896, swelling its total fleet to more than 260 vessels with 2,500 employees. Within three years it was in financial trouble because sail could no longer compete with steam, and hundreds of men had to find work in steam trawlers sailing from other ports. The final blow was an explosion in the company's engineering works at Barking in 1901, and the bulk of its vessels had to be sold to meet claims arising from the accident.

Fleeting was discontinued by Grimsby vessels in 1901 but was continued from Hull by four fleets – Red Cross, Hellyers, Gamecock and Great Northern. Carriers now had less trouble finding the fleet because it made so much smoke. Each fleet sailed under the nominal control of an admiral, as before, and an old smack was used as a mark boat. Steam trawlers carried no ice because fish were delivered daily to the cutter but this allowed them to carry very large supplies of coal to last them three or four weeks. In 1936 deliveries to Billingsgate by sea were stopped because of congestion and the two remaining fleets, Gamecock and Red Cross, transferred to Fleetwood where they merged with what remained of the Short Blues to form a new company.

Until the First World War, when 249 were sunk by enemy action, sailing smacks based at a small number of other ports around the coast managed to survive by trawling inshore waters such as the Straits of Dover and the Bristol Channel for soles, which were the most valuable fish. In 1914 there were still 610 smacks operating – 193 at Brixham, 152 at Ramsgate, and 265 at Lowestoft. Yarmouth suffered most from the decline of sail, its 400 smacks in 1889 being reduced in ten years to a handful of aged and obsolete boats.

While trawling at Yarmouth withered to nothing (although its autumn herring season was booming and attracted hundreds of drifters), Lowestoft's landings increased by two and a half times during the ten years ending in 1902. Its trawler dock was at the harbour mouth, not a mile up the river as at Yarmouth, and its railway facilities were handy to the quays. In 1904 it had 300 trawlers of which about fifty were steamers, plus the same number of drifters and other boats in shrimping and line-fishing.

Unlike sailing smacks the steam trawlers could work regardless of wind and tide and were prevented from fishing only by severe weather. They could fish all day and every day. A fisherman's life was hard and cruel by any standard, now steam had brought the tread of the mill into trawling: the drudging repetition and limited horizons of the factory production line as well as its smells of oil and the clatter of machinery. There may have been little to occupy the mind during periods of enforced idleness due to summer calms in the sail-trawling era, but the pace of life was more acceptable and a man was at least compensated for tedium and harsh conditions by scope for self-improvement.

The new mood of profitability and persistence in the company-owned fleets was reflected in the names of trawlers. In the more leisured days of sail all the human virtues and some of their frailties were represented – *Wayside Flower, Eye of Providence, Star of Hope, Brotherly Love, Happy Return, I'll Try, Fear Not, Fame, Glory*; at Grimsby a smack was named *Robert the Devil* after a racehorse, and one owner built the *Coventry* which was followed by *Lady Godiva* and

The diagram illustrates the procedure for boarding fish that was laid down to prevent accidents after steam was introduced to the trawling fleets.

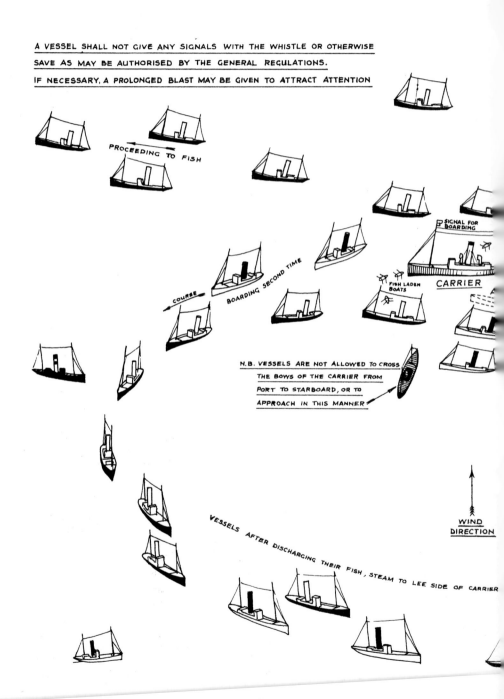

A VESSEL SHALL NOT GIVE ANY SIGNALS WITH THE WHISTLE OR OTHERWISE SAVE AS MAY BE AUTHORISED BY THE GENERAL REGULATIONS.

IF NECESSARY, A PROLONGED BLAST MAY BE GIVEN TO ATTRACT ATTENTION

PROCEEDING TO FISH

SIGNAL FOR BOARDING

BOARDING SECOND TIME

COURSE

FISH LADEN BOATS

CARRIER

N.B. VESSELS ARE NOT ALLOWED TO CROSS THE BOWS OF THE CARRIER FROM PORT TO STARBOARD, OR TO APPROACH IN THIS MANNER

VESSELS AFTER DISCHARGING THEIR FISH, STEAM TO LEE SIDE OF CARRIER

WIND DIRECTION

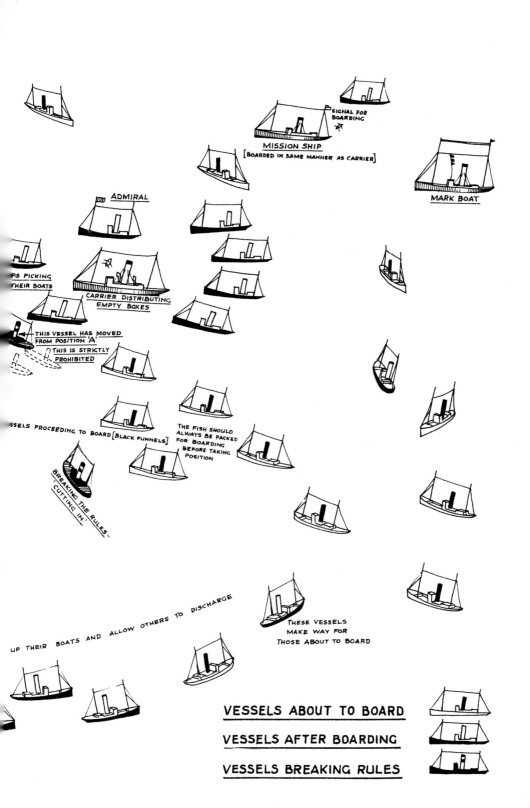

SIGNAL FOR BOARDING

MISSION SHIP
[BOARDED IN SAME MANNER AS CARRIER]

MARK BOAT

ADMIRAL

PS PICKING
THEIR BOATS

CARRIER DISTRIBUTING
EMPTY BOXES

THIS VESSEL HAS MOVED
FROM POSITION 'A'
THIS IS STRICTLY
PROHIBITED

SSELS PROCEEDING TO BOARD [BLACK FUNNELS]

THE FISH SHOULD
ALWAYS BE PACKED
FOR BOARDING
BEFORE TAKING
POSITION

BREAKING THE RULES
'CUTTING IN'

UP THEIR BOATS AND ALLOW OTHERS TO DISCHARGE

THESE VESSELS
MAKE WAY FOR
THOSE ABOUT TO BOARD

VESSELS ABOUT TO BOARD

VESSELS AFTER BOARDING

VESSELS BREAKING RULES

Peeping Tom. But steam trawlers bore names like *Gleaner, Zealot, Breadwinner,* or just numbers, and house flags gave way to funnel markings. The profusion of new companies could lead to complications. There is a story that when King George V and Queen Mary were on a tour of Grimsby fish dock and the assembled steam trawlers he asked an alderman where the urinal was.

'Can't say offhand, sir,' he was told. 'Do you know what she's got on her funnel?'

Intensive fishing by increasing numbers of bigger and more powerful vessels soon caused a depletion of fish. In ten years the average trawler catch in the North Sea declined by nearly a quarter, from 18.64 cwt in 1903 to 14.08 cwt in 1913. The annual operating cost of one trawler was about one thousand pounds, and owners found themselves having to pay ever-increasing expenses for diminishing returns. As the North Sea became a desert new grounds were found further from home, but in travelling further, at greater cost, trawlers returned with fish that earned the lowest prices because it had been for so long on ice.

In 1901, when Grimsby was straining every nerve to maintain its fantastic rate of expansion, trawler owners, looking for ways of reducing their capital outlay, decided to reduce the scale of wages for engineers and deckhands, and make up the difference by sharing the profit.

The profit, or 'clear money', was calculated after expenses for coal, ice, food, engine had been deducted, plus five per cent, from the sale of the catch. It was divided into fourteen shares, of which the skipper received 'one share and one quarter and one half-quarter' and the mate one and one-eighth. An average settling bill after eight weeks was about £500, expenses about £180. The skipper therefore earned a little more than six pounds a week, the mate about five pounds, out of which they provided their own food at retail prices. A chief engineer earned 46s. a week, which the owners wanted to change to 30s. plus 4d. for every £1 clear money; after an average trip he would collect 47s. a week. Earnings could be considerably higher, or a lot lower, and neither the engineers nor the deckhands wanted to take the risk so they went on strike. The owners, who had chosen the summer slack season in which to force the issue, imposed a complete lock-out. It lasted fourteen weeks.

At first the men behaved with exemplary moderation in what was described as a local and national calamity. The whole town was paralysed. 400 steam trawlers lay idle in the docks. Women and children starved or lived on charity. Then rioting trawlermen looted

64. *Sails drawing, helm a-lee as the tow-rope is cast off, the Lowestoft trawling smack* Francis Robert *departs for the North Sea fishing grounds while a steam drifter makes for the harbour entrance.*

and set fire to the offices of the Federated Owners' Protecting Society, its secretary and staff only just escaping the mob. 300 extra constables were drafted in from Sheffield and Manchester. The Riot Act was read in the streets.

Hull cashed in on the stoppage and a handful of old smacks at Grimsby made vast profits sailing their catches direct to Billingsgate. Finally the men agreed to arbitration and compromise was reached, a chief engineer receiving 34s. a week plus 3½d. which, after average fishing, would total about 54s. a week.

The search for new fishing grounds had started in 1891 when three trawlers landed fish from Icelandic waters, but the price had fallen and the experiment was not repeated until 1902. Meanwhile trawlers had fished the Bay of Biscay in 1893 and the Faroe Islands in 1898. In 1904 the rich fishing grounds off the Murmansk coast of north Russia were discovered, and trawlers ventured south to the coast of Morocco and west of Ireland, to lonely Rockall. Just prior to the outbreak of war the Grimsby trawler *Itonian*, on her maiden voyage, found a good haul near Jan Mayen Island far north of Iceland, and subsequently new grounds were discovered at Bear Island, Spitsbergen, Greenland and Labrador.

Unfortunately few details of these pioneering voyages are known. It is romantic to think now of hundreds of rusty, battered, scruffy little British trawlers with tall, thin funnels exploring every fishing ground in the north and east Atlantic between the high Arctic and the tropics. At times whole crews were arrested for poaching in territorial waters from Morocco to Murmansk. At the time, however, it was a job of work – a question of survival rather than adventure. When a new ground was discovered the skipper told his owner who told his other

THE END OF SAIL

First class smacks registered:

	1887	1903		1887	1903
Hull	448	0	Dover	24	0
Grimsby	839	34	Rye	32	0
Scarborough	90	20	Brixham	159	193
Great Yarmouth	627	86	Plymouth	89	50
Lowestoft	376	356	Hoylake	34	0
London	105	0	Fleetwood	43	43
Ramsgate	140	138			

skippers, but secrets were difficult to keep in the dock-side pubs and the news soon spread.

In Icelandic waters skippers found prolific and seemingly inexhaustible grounds and by the early 1930s more than a quarter of Britain's white fish and about half her cod came from the waters of that fiercely cold and ruggedly barren shoreline. The irony was that the

HOW THE FISHING GROUNDS CHANGED

| | Tons of white fish landed in England and Wales | | |
	North Sea	Beyond North Sea	Total
1903	260,000	68,000	328,000
1904	231,000	78,000	309,000
1905	207,000	93,000	300,000
1906	218,000	204,000	422,000
1907*	221,000	224,000	445,000
1908	204,000	228,000	432,000
1909	199,000	241,000	440,000
1910	186,000	241,000	427,000
1911	186,000	252,000	438,000
1912	187,000	250,000	437,000
1913	169,000	248,000	417,000

Peak year: catches approximately equal

LANDINGS OF FISH BY STEAM TRAWLERS, c. 1902

| In kits – one kit about 120 pounds | | | |
From North Sea		From Icelandic waters	
Springfield	225	Rome	1,576
Iolanthe	243	Plutarch	1,503
Argonaut	222	Othello	1,142
Sabrina	209	Norman	1,026
Honoria	208	Titania	792

fish brought home from such distant grounds, though carefully packed in ice, was of such low quality that the old sailing smack skippers would have chucked it over the side as sub-standard, or sold it at rock-bottom prices as offal. However, there were 25,000 fish and chip shops in Britain before the First World War, where customers were less exacting than fishmongers, and it was the search for quantity rather than quality, to keep that market supplied, that was the driving force behind steam trawling.

Cement, Iron-rust and God's Mercy.

A steam trawler before the First World War was typically about 120 feet long and 20 feet wide. The reciprocating, triple-expansion engine, of 50 or 60 horsepower, was installed about two-thirds of the ship's length back from the bow, which put the centre of gravity well aft with the result that the bows pitched a great deal more than the stern. Living conditions in the foc'sle – the deckhead hung with dozens of pairs of wet socks and illumination coming from a smoky paraffin lamp – was only one degree better than the chain locker. When driving into a head sea the pitching was so excessive that mattresses and anything not tied down were liable to be left in mid-air as the bows of the ship dropped. Then, as they began to fall, they hit the deck which was surging upwards again. If you happened to be lying on the mattress the bone-jarring thump was not quickly forgotten.

In following seas the bows tended to bury into the water, and in those days there was no 'whaleback' or arched shelter over the bows, so water continuously poured into the foc'sle. At least the eight men who lived there were spared the thunder of the screw turning a few inches from their heads, which was suffered by the skipper, mate, and two engineers who lived in the saloon right aft. But the foc'sle men had to lash down the anchor chain in its hawsepipe to stop its sledge-hammer clanging. The early steam trawlers were cold, roughly built, and uncomfortable. They were so ramshackle that one fisherman remarked, 'These puddle-jumpers are held together by cement, iron-rust, and God's mercy.'

Trawlers soon inherited the bridge deck that had become standard in merchant ships and even the unheard-of comfort (for a fishing boat) of a wheelhouse. Many trawlers built around the turn of the century, particularly those for Hull owners, had the bridge positioned right aft and some had the wheel position off-set to give visibility past the high 'Woodbine' funnel. Some of these 'bridge aft-side' ships remained in operation up to the Second World War, but they were hard work for the trimmers whose job was to stoke the boilers. The coal bunkers were forward of the engine and the boilers were aft, so coal had to be carried in baskets; but first the two wing bunkers on either side of the engine had to be cleared and these were so narrow that the stokers had to throw tons of coal backwards over their heads.

Outward-bound trawlers were so laden with coal that they barely had positive buoyancy. Reserve coal carried on deck was used first, being tossed down the stokehold ventilator as required, then the reserve bunker was emptied so it could be cleaned out and used to store fish. Being low in the water the trawlers were very wet and frequently awash, necessitating hasty action by the crew to throw overboard heaps of coal, or fish, if the scuppers became jammed. 'We dived at Spurn and came up in the fleet,' was how one skipper described a wet trip.

A trawler's ability to roll away from an approaching wave and so protect its crew working on the open deck was important for safety, but despite small steadying sails this made them much less comfortable than the sailing smacks. Sanitary facilities of a kind did exist: there was a sentry-box lavatory on deck for'ard – which was usually put to better use by the mate as a shackle store – and another right aft which was never used because the pitch and roll of the ship sent columns of cold sea water gushing upwards through the hole. Instead, men squatted on the small-coal in the stokehold then shovelled it into one of the boilers.

In the first steam trawlers skippers napped on the deck of the wheelhouse so they could be woken with a kick when needed, but in later ships the wheelhouse was built for'ard of the engine room with room for a combined chart-room and skipper's cabin beneath. Just forward of the wheelhouse was the powerful double-barrel steam winch on which were wound two steel-wire warps 1,000 to 1,500 fathoms in length that were used to tow the trawl. Automatic guiding gear was not introduced until the 1920s and men had to guide the warps on to the drums with big iron 'toggle' bars.

The winch introduced a new dimension of horror for careless deck-

65. *'Cement, iron-rust, and God's Mercy' is what one fishermen said kept the early steam trawlers afloat. Here the Aberdeen steam trawler* Strath-devon *heads for home.*
66. *Stokehold of a steam trawler: the steam age lasted only a man's working lifetime but it changed the face of the fishing industry and made it a tread-mill, capital-oriented operation.*

hands. In 1897 the third hand of the *Asia* was taking off his coat near the winch when it was caught in the machinery and drew him in: the body was instantly decapitated and the head rolled down the deck. A deckhand of the *Talifoo* was coiled round the drum of the winch while guiding the warp, and by the time the winch was stopped his body was coming out feet first – 'The sight that met the gaze of the crew was a horrible one,' reported a Grimsby newspaper. Stopping the winches quickly was always difficult: one trawler, the *Hildina*, winched herself right under when her trawl came fast on an obstruction.

Greater power allowed improved trawls to be used. The first development was the 'otter' trawl which dispensed with the heavy wooden beam and used heavy iron-shod wooden doors, acting like vanes and sheering out from either side, to keep the mouth of the trawl open. These doors, or otters, were adapted from a now illegal method of catching pike and trout on Irish loughs and were in general use throughout the steam fleet before the turn of the century.

Another important modification was adapted from a French idea in 1922 when the length of the bridle was increased to about 100 yards and longer wing nets were fitted. The wide spread of the bridles and wing nets of this Vigneron-Dahl or 'V.D.' trawl had the effect of startling fish towards the centre where they were scooped up by the main net, and catches of haddock increased by fifty per cent. Also, large round bobbins were fitted along the foot-rope so the trawl could travel over hilly ground. With minor modifications this system has remained in use up to the present day.

Shooting and hauling big trawl nets with twin warps and enormous doors weighing more than a ton apiece was a tricky operation, especially at night with the deck rolling and pitching and slippery with ice, and sleet and spray driving into your face. A number of things could go wrong: the ship could get over her gear and foul the propellor, the tide could turn the trawl net inside out, the doors tended to sheer about like kites and could cross each other or charge into the mouth of the net.

The net was always worked from the windward side of the vessel so it did not go under the keel as the ship made leeway. The warps went over the side by way of arched steel supports called 'gallows', one aft and the other for'ard. After the net was shot the length of the warps was adjusted with the winch to ensure that the trawl came square-on. When towing, the two warps were pulled into the side of the ship by means of a special block and tackle.

Before the First World War deckhands worked by the light of 'moonboxes' which were made on board by twisting oily waste in wire to make a kind of flare. These were stuck into the deck stanchions and burned for a couple of hours. Later there were acetylene lights hung on a boom along the deck and fed by a pipe from the engine room. Heavy

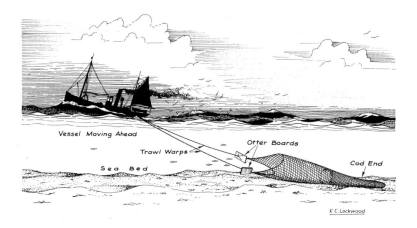

Labels on illustration:
Vessel Moving Ahead
Otter Boards
Trawl Warps
Sea Bed
Cod End
K.C.Lockwood

Trawler. A trawler tows a bag-like net along the surface of the sea-bed; fish are stirred up by the long warps and frightened into the mouth of the net.

rolling put the lights out and the pipe often filled with water; explosions of gas were commonplace. Rough handling of acetylene drums caused several fatal accidents and in 1922 the trawler *Betty Johnson* blew up as she was leaving the dock with a bang that shook Fleetwood.

On three-week voyages to Arctic waters life was not as unremittingly boring as it had been – and still was – in the North Sea fleets. But for short periods, while actually fishing, conditions were a great deal tougher and the work more intense. Steaming to and from the fishing grounds, four or five days each way, the crew could relax, working reasonable hours while assembling or stowing the fishing gear. Once fishing started it did not stop until the holds were full, the weather broke, or coal ran low. Men worked thirty hours at a stretch and were lucky to get four off. Gutting and hauling was more or less continuous, with no let-up. Coming into the warmth after hours on deck men were unable to roll cigarettes and fell asleep with their faces in their food. Skipper Jack Ellis of Hull, who first went to sea in 1919 and was a skipper for many years, remembers a hapless deckie-learner who fell asleep on the lavatory and the crew filled his trousers with cod livers.

Gloves were frowned upon by some skippers because they reckoned it slowed the gutting. With no mission smack cruising nearby to provide knitted cuffs, the men protected their wrists against chafe by sliding women's stockings on their forearms. Hands became tough and horny, and had to be softened with carbolic acid. A man doing an intricate job on deck, such as tying a knot, knew of only one quick way

261

to warm his fingers – he asked another deckhand to urinate on them. Spawning haddocks sucked up sand which stuck to men's fingers, rubbing them red-raw and causing a painful rash: one skipper can remember having to steer the ship with his wrists because he could not bear to bend his fingers round the spokes of the wheel.

Crews lived well, better than in smacks where the cooking was at the mercy of an untrained, brow-beaten and seasick lad. Steam trawlers carried full-time cooks. Meals were dished up in the saloon which, in contrast to the spartan foc'sle, was panelled in mahogany. When the ship sailed the cook put the meat on ice in the fish-room then prepared his own brine and pickled it in a cask. Fresh vegetables were carried in the lifeboat, which was handy to the galley on the aft deck, and if the cook lost some potatoes he was likely to find them in a ventilator where the crew had put them to keep out the draughts. Most men brought their own reserves of sugar and tinned milk in case supplies ran out, as they often did.

At sea trawlermen wore fearnought trousers – 'Bullet-proof, never mind water-proof' – a big oil frock that reached to the ankles, and a towel round the neck. It did not matter how wet your top half became, but if your feet were wet you were allowed to leave the deck to change: skippers realised that a man could not work for long if his feet were wet and cold. Skippers got even less rest than the crew and at meal-times dashed below to bolt three courses in less than twenty minutes. They were vulnerable to a new occupational disease: phlebitis, or inflammation of the arteries, mainly in the legs, from long hours of standing on the bridge.

Despite the greater distances steamed to and from the fishing grounds navigation was still largely by rule of thumb. By the Merchant Shipping Acts of 1894 and 1904 skipper and mate of fishing boats larger than twenty-five tons were required to have official certificates of competency for which candidates were examined by the Board of Trade. These covered rule of the road, use of chart, compass and lead-line, and general subjects such as distress signals and how to rig a jury rudder. It is an indication of the educational standard expected that Reed's Guide to Skipper and Mate Examinations, published in 1904, had the twelve-times multiplication table as a frontispiece. Extra certificates of competency were required by the insurance companies which set their

67. *Ships get big, voyages longer, costs higher, and the work remains as hard as ever: even in modern vessels fishing is the business of hauling in nets.*

68. *A small haul for this Scottish steam trawler soon after the First World War as the cod-end is swung aboard and the mate (left) prepares to slip the knot that will let the fish drop into the pounds on deck.*

own papers that included simple arithmetic, elementary use of the sextant, and questions such as:

The wind is on the starboard beam: suppose you see a green light two points on the lee bow, what would you do?

Answer: *Keep clear, the other ship is close-hauled on the port tack.*

Nearly all questions were oral. Men who had already been at sea were able to avoid the small number of written questions, and several of the most successful skippers in the 1920s had all the necessary certificates but were unable to read or write.

Fishermen's charts were printed with the compass courses to steer between the main landmarks and these were not difficult to remember. Navigation was by dead reckoning. Hand-lead and compass had been used by generations of fishermen to feel their way over the grounds of the North Sea and few men saw any reason to change their ways. If a skipper was seen using a sextant the crew were liable to lose faith in him – 'The old man knows bugger all.'

In the light of this traditionally casual approach to navigation and ship-handling it is hardly surprising that strandings, collisions, founderings and other accidents still continued to occur at much the same rate as in the days of sail. In the first half of 1904 no less than twenty-one British trawlers were lost at sea. One skipper had a lucky escape off Flamborough Head when his trawler was sunk by an oil tanker and he was able to leap from his wheelhouse on to the other vessel's anchor, carrying his helmsman in his arms. When the trawler *Chindwin* hit a rocky part of Scotland in the winter of 1895 the crew lassooed a pinnacle and hauled themselves ninety feet to shore then, battling against frozen spray and snowdrift, climbed a precipice and found themselves on a lonely, snow-covered moor; after being rescued by shepherds nearly every man had to suffer an amputation of toe, finger or foot because of frostbite.

In Icelandic waters storm, frost, icebergs, and long hours of darkness, and a low-lying coast so remote from habitation that special shelters and caches were built for the aid of shipwrecked fishermen, were not the only hazards of fishing. Trawlers were even then running the gamut of Icelandic gun-boats protecting their three-mile limit, and the long nets set on the sea-bed for cod by fleets of local motor boats called 'snibbies'. By lining up different headlands and other coastal features, and through long experience, skippers built up their own sets of 'marks' where they found the best fish. Said one skipper: 'You would have one trawl door scraping the rocks and the other right on the three-mile limit, and a gun-boat right astern knowing that you knew your business but waiting for you to make just one little mistake – they were always scrupulously fair but kept a very close eye on you.'

One skipper poaching in an Icelandic fiord was caught red-handed by a patrol boat and an officer came aboard saying 'I am the law and

I am going to arrest you for illegal fishing.' But the trawler skipper said 'Oh no, I'm going to arrest you!' and took him back to Hull. The skipper never dared to return to Icelandic waters in case he was arrested for abduction as well as poaching.

Nearer home, the whole country was thrown into a state of dangerous excitement and war fever when trawlers of the Gamecock fleet were fired on by the Russian navy on the Dogger Bank during the night of October 21–22, 1904. On that fine, hazy night thirty trawlers, including two Mission hospital steamers with their gear down, were working in twenty-three fathoms of water when the Russian fleet appeared over the horizon. This was one of the biggest naval fleets ever assembled, and it was on its way to annihililation at the hands of the Japanese. The first section passed safely through the little fishing fleet. The second squadron, of four battleships, steamed just across the bows of the trawlers, playing their searchlights.

Then, as Walter Wood describes it in *North Sea Fishers and Fighters*, a bugle rang out in the night and the great battleships instantly opened fire with heavy guns and machine guns. The Russians had received reports that the fleet would be waylaid and attacked by Japanese torpedo boats which had succeeded in secretly reaching the North Sea. The trawlers, anchored in position by their fishing gear, were peppered with shell-fire but the ill-trained Russian gunners were wildly inaccurate and nearly paralysed with fright. The trawler *Crane* was sunk, her skipper and bosun killed and five wounded, and five other trawlers were damaged by shot. When it was reported in Hull forty hours later by the shot-riddled *Mino* there was such a general clamour against the Russians that the Czar promised 'complete satisfaction' and after an international commission sat in Paris the Russians paid out £65,000 compensation. For the trawlermen the 'Russian outrage' was a fore-taste of the war that was to come only ten years later.

THE BLACK MONDAY GALE (December 1883)

As told by the skipper of a steam trawler and published in Toilers of the Deep, *1893*

The storm began with a blinding downpour and a gentle puff of wind at intervals, each one stronger than the last, until at last it rose and increased in fury to a hurricane; and then, in a moment, it dropped to a dead calm without any warning. The wind had been from the south-west in the squalls and our aneroid was rushing up at an alarming rate. This, coupled with the sudden dropping of the wind, was a very bad sign so we hove up our fishing nets and prepared for a 'blow hard'. We might have steamed out of danger if we had known what was coming,

but we had reaped a rich harvest on that same ground the night previous and were loath to leave it. But when an hour had gone, and no sign of wind, we began to hope the worst had passed, and were actually preparing our nets again when we heard a low rumbling sound, like distant thunder, approaching from the nor'-nor'-west.

On looking in that direction we could see clouds piling themselves up almost as quickly as thick black smoke from a furnace. Nearer and yet nearer on they came, and in about two minutes the gale in all its fury was upon us, screeching, roaring and howling, bearing our vessel over almost at once on her broadside; whilst the wind was so strong the sea was comparatively smooth. What we had to fear now was the wind dropping. This happened towards tea-time then up rose the 'sea lions', roaring, frothing and leaping in savage fury at our devoted ship. The first huge mountain of water broke on board whilst we were at tea, washing the man at the wheel off the bridge, entangling his legs in the wheel chain and seriously injuring him. We rushed on deck and succeeded in getting him clear, and carried him below.

Darker grew the night, higher rose the sea – rushing and leaping in mad fury, threatening at every moment to engulf our vessel as she lay, a mere toy, upon that wild waste of waters. Sleep, of course, was out of the question and we remained fully dressed, dreading the worst, yet trying to persuade each other that we were hoping for the best. About 11 p.m. the second great mountain of water broke on board. We heard the watch on deck shouting 'Water!' and springing up the companion hatchway I heard the words, 'Lord Save Us!' and 'Oh my poor wife and children!' We were all on deck by this and saw that awful sea rushing at us. I have read that, on coasts, where earthquakes are frequent, the sea sometimes leaves the shore for miles and then, gathering force, comes rushing back in one mountainous wave. Here now was one in the middle of the ocean, stretching away across the wind as far as the eye could reach. We watched it gathering force. We could hear the crashing grinding noise as it rushed at us and not one man on board but thought his last moment had come.

We felt our poor vessel tremble with the vibration of that thundering sea, towering as it seemed to us some twenty feet above us. Our ship seemed to be drawn broadside-on right underneath it, heeling over towards it, and then it fell. In a moment we were buried fathoms deep. How can I describe the next few moments? Screams of mercy, cries for help, ejaculatory prayers, soul-stirring indescribable sounds, the

---

69. *Sextants were rare in fishing boats, even in distant-water trawlers which fished in Arctic waters; this Hull skipper is wearing the traditional fisherman's gansey and is loyal to the old-fashioned drop-front serge trousers.*

artillery-like cracking of the torn sails, howling wind and roaring sea. It was God's mercy that we ever came to the surface again. But what a change! Masts, funnel, ventilators, lifeboat, trawl beams, nets, bridge, compass, companion hatchway – *all* had gone; but worse than this, our cabin was full of water and there was four feet of water in the engine room. Our pumps were choked with small-coal and would not work. We were half drowned, bruised, and bleeding, and totally unfit to cope with this danger, knowing, too, that with one more sea like the last all would be over. And this was likely to happen at any moment.

However, life is sweet and the fear of death coupled with hope – truly the sailor's sheet anchor – gave us courage to fight the raging elements above and below us, and taking off our seaboots and oilskin caps we began our battle for life. It was bale, bale for eight hours without one moment's rest, but the strain mentally and physically was telling, and with two feet of water yet in the cabin we could do no more. Thoroughly exhausted we sat on the lockers and more than half dead we watched our clothes and bedding floating around us. All that day our vessel lay in the trough of the sea, at the mercy of the wind and waves. Four times that night we were hove down on our beam ends and each time we did not expect she would right again. Higher and higher rose the water in the cabin, darker grew the night, louder roared the sea, without food or fire, bitterly cold, up to the armpits in water as we sat, worn with toil, anxiety and fear, expecting every moment the dread summons to appear 'across the river'. I was twenty-eight years old which was about the average of each one of us, and it seemed hard to die in the strength of our young manhood, with health unimpaired and our loved ones would never know how we had sat thinking our last thoughts of them.

I think it would be about four or five o'clock next morning I was startled out of a stupor into which we had all fallen by the extra heavy rolling of our vessel and the water rushing madly about the cabin. I felt sure this was a good sign, and ventured to peep on the deck. Oh how glorious! The sky was perfectly clear and fairly studded with beautiful stars. The sea was as turbulent as ever but its fangs were drawn, its power to harm was fast going away. After praying, we found some tobacco and matches in the foc'sle – one smoke was all we could get in the shape of refreshment. After our smoke we set to work to clear the wreck but it was not until fourteen more hours' hard work that we were able to get steam up and steam slowly home.

Not Fish Enough.

In the summer of 1883 the fishing industry demonstrated its dynamism and prosperity with a spectacular International Fisheries Exhibition held at South Kensington, London. More than 1,500 companies involved in every aspect of fishing, from trawler owners to hook makers, took part. Thirty-six countries and colonies were represented. Two million people visited. Fishermen could see nothing for themselves but a golden future.

In his inaugural address the eminent naturalist and scientist Professor T. H. Huxley, as president of the Royal Society, spoke words that must have sent a thrill of excitement and confidence through every man with money in this booming industry: 'I believe it may be affirmed with confidence that in relation to our present modes of fishing a number of the most important sea fisheries, such as the cod fishery, the herring fishery, and the mackerel fishery, are *inexhaustible*; that is to say, nothing we do seriously affects the number of fish.'

Huxley based his judgement on salmon fishing, in which five per cent of the stock was caught every year. If he had looked at the decline in catches of plaice in the North Sea, to name just one of many species, his note would have been far from confident. The average annual catches of plaice by four Grimsby smacks fell from 998 cwt in 1867 to 291 cwt in 1880 and 168 cwt in 1892. These figures refer to *sail*: the effects of large and powerful steam trawlers were yet to be felt.

In 1835 the total width of sea-bed that could simultaneously be swept by the trawls of the combined fleets of smacks in the North Sea was reckoned at less than a mile. In 1863 it was about eight miles, and in 1883 more than twenty miles. In 1903, with fewer vessels but wider nets, the total extent of nets had not greatly increased but frequency had nearly tripled. Now the twenty miles of trawl swept the North Sea for greater distances, not once a day but three or four times a day.

Although total landings increased as vessels multiplied and were driven further and further from home to seek new grounds, it was not long before the distant waters too began to suffer from the effects of the same disaster that had hit the North Sea – over fishing. Fishermen could not be content. The sea did not have fish enough.

Since the discovery of the Silver Pits in about 1843, or for that matter since Thames fishermen petitioned the king about the 'wondry-choun' five hundred years before, trawling itself had always been the bone of contention between fishermen. The clamour became so intense that in 1863 a Royal Commission, of which Professor Huxley was a member, took evidence from fishermen in eighty-six ports. Line fishermen argued that a fish must be in good condition while it was feeding, but a trawl swept up every fish regardless of its condition; trawlermen contended that a hungry fish was a lean fish, and that trawled fish were too well-fed to take baits. Some complained that trawling damaged the sea-bed; trawlermen said it did the fish good – 'It turns up worms and slugs, and fish follow the net like crows after the harrow.' Inshore fishermen claimed trawlers had completely ruined the east coast grounds, and fish were not only scarcer but smaller. Trawlers were accused of scooping up tons of herring spawn and selling it for manure, which they flatly denied.

The commission cleared beam trawls as not wasteful, concluded that complaints among different classes of fishermen were nothing to worry about, and recommended that the 'complicated, confused and un-satisfactory' laws relating to open-sea fishing be repealed to permit 'unrestricted freedom of fishing'. The result was that for nearly seventy years, until the privilege of freedom was withdrawn and the government assumed wide powers of direction in all fishing matters, fishermen on the high seas could fish how, when and where they liked.

The agitation against trawling did not stop however, and in 1878 another inquiry was conducted, this time by the Home Department. After visiting forty-eight ports the board found that there was no wasteful destruction of fish, the fish supply was not decreasing, and damage caused by trawlers to drift-nets and lines was only accidental and due mainly to defective regulations for the use of lights at night.

Herring fishermen's complaints against trawlermen were seen in a new light when it was pointed out that cod caught in the North Sea every year ate twice as many herrings as were caught by British drift nets, and the gannet colony at St Kilda alone accounted for as many herrings again. One scientist estimated that there were two million gulls in the United Kingdom, and each ate two hundred herring fry during the two-month season; these 34,000 million herring would be worth £24 million if caught. Thus he advocated destroying all gulls since each one cost the country twelve pounds a year.

---

70. A basket of iced cod is swung out of the hold of a trawler in Grimsby fish dock and will be hauled ashore by the "bobber".

The point of these fanciful propositions is that nobody really knew. The study of marine zoology was in its infancy, as was the science of oceanography which had been established with the three-year circumnavigational research voyage by *HMS Challenger* in 1872–5. In 1884 the Marine Biological Association was formed to study the marine environment and fish, and four years later it opened a research station on Plymouth Hoe. While the reconstituted Fishery Board for Scotland began sponsoring scientific investigations in 1882, there were no funds available in England until 1902 when the government established a new laboratory at Lowestoft which was equipped with a trawler for sampling catches. Later the Marine Biological Association was commissioned to continue research it had begun on its own initiative but stopped for lack of funds.

These efforts were stimulated mainly by pressure from abroad. In 1899 the King of Sweden called a meeting of European nations with a stake in the North Sea. This led to the formation in 1901 of the International Council for the Exploration of the Sea (I.C.E.S.) and a period of international co-operation between eight (later fifteen) countries that has been much expanded in scope and has continued since. Seas were mapped out for investigation and each country agreed to set up its own research department.

Britain's first research vessel, which operated from Lowestoft from 1902 until 1910 when a converted minesweeper was purchased, was a modified Grimsby steam trawler appropriately renamed *Huxley*. A

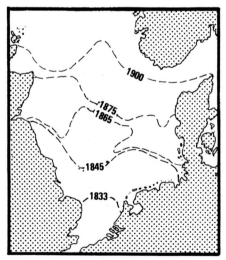

SPREAD OF ENGLISH TRAWLING IN THE NORTH SEA DURING THE 19TH CENTURY

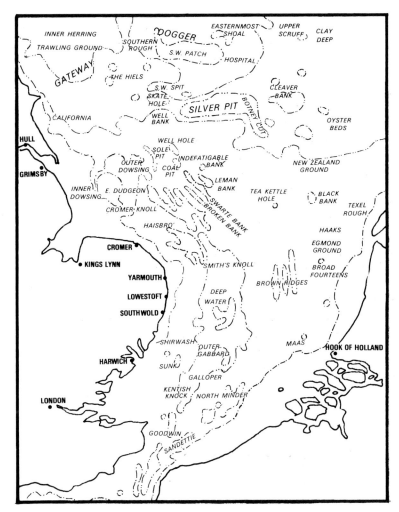

THE FISHING GROUNDS OF THE SOUTHERN NORTH SEA

beam trawl was carried on the starboard side and an otter trawl on the port side, allowing sample catches to be made on grounds used by steamers or smacks. Work could now begin on conservation. It was conservation not for the benefit of the fish – but for protecting productivity and resources for the sake of Man.

When the first experiments in tagging plaice showed how rapidly the grounds were being depleted those who advocated fishing restrictions to conserve fish stocks were at last able to base arguments on fact rather than theory and conjecture. About 54,000 tons of this fish were taken out of the North Sea by European fishermen in 1908. Sir Alister Hardy, one of the country's first fisheries scientists, estimated that England's landings of 215,183 cwt of 'small' plaice at East Coast

273

ports numbered 100 million fish, but to catch this number probably another 400 million unmarketable and under-sized plaice had been destroyed. The pressure on plaice was demonstrated when the *Huxley* trawled hundreds of plaice which were tagged with special buttons fixed with silver wire and released. Within twelve months more than forty per cent had been caught again, and in some experiments as much as seventy per cent.

There were also plans to transplant millions of baby plaice from over-crowded nursery grounds near Holland to the Dogger Bank where there was plenty of food. They grew six times more quickly, and their value increased up to ten times. But as nothing was done about restricting fishing grounds, because nations could not agree on which parts of the North Sea should be closed, the ambitious transplanting scheme lapsed because there was no guarantee that if Britain did the work her own fishermen would benefit.

Ultimately, the case for conservation was proven by the First World War. Minefields and naval operations brought North Sea fishing almost to a complete stop for four years. When fishing was resumed in 1919 the average landing per trawler for every day's absence from port more than doubled – from 14.3 cwt in 1913 to 30.6 cwt, giving scientists the most vivid demonstration of how a fishing ground could recover. Prices had also rocketed (by 230 per cent) during war-time and the result was a mad scramble to cash in on the bonanza. Within four or five years the North Sea had less fish than before. Trawlers barely paid their expenses which had quadrupled, and fishermen faced the marine equivalent of the North American dustbowl.

This also had a crippling effect on inshore fishing, the administration of which had been put on a more formal footing by the Fisheries Regulation Act of 1888, which legislated for such things as close seasons and mesh sizes. Scotland had its own fisheries board.

Pipe Stalkies.

Between 1900 and the First World War herring fishing in the North Sea reached its peak of prosperity. During the November 'herring moon' more than 3,000 miles of nets were shot every night off the coast of East Anglia. Herrings were landed by the hundred million, and in terms of sheer numbers of fish handled Great Yarmouth and Lowestoft became (during the season) the busiest fishing ports in the world. In 1913, the most successful season ever, 854 million herring worth nearly £1 million were landed at Great Yarmouth in fourteen weeks, and 6,000 Scottish lassies slaved away on the denes amid mountains of barrels to gut and cure them; another 436 million were landed at Lowestoft.

During these years 1,800 steam drifters were built: little ships about 90 feet long with a beam of nineteen feet and nine feet of depth in the hold. As with trawlers the distinctive design was common to both England and Scotland. Voyages were comparatively short so there were bunkers for only fifteen tons of coal. The triple expansion engine gave a service speed of about nine knots: but when racing back to port with a good catch to get a quayside berth in order to obtain the best prices, speed could be increased at great risk by putting a heavy spanner on the boiler safety valves.

From the straight stem to the low and deep counter stern the deck had a graceful sheer and there was about three feet of freeboard, with low bulwarks, amidships. The steel engine casing was characteristically rounded over at the edges and from it, behind the 'telephone kiosk' wheelhouse, sprouted a tall thin funnel from which the early steam drifters were known by Scottish crews as pipe stalkies. The fish hold forward had space for the drift nets – up to 130 were carried – and 250 crans of herrings, and it was covered by a very large hatch.

To keep the drifter's high bows into the wind while lying to her train of nets the foremast was lowered into a chock on the wheelhouse roof and a steadying sail was set on the mizzen. The foremast was raised only in port, where it served as a support for the cranning pole, or

71. *Congestion off Lowestoft pier-head, probably just before the First World War as (left) two trawling smacks are towed seawards by a paddle tug (concealed), a Fifie (right) lies head-to-wind as her crew prepare sails for hoisting and wisps of steam blow from the capstan engine in her stern; a rolling "pipe stalkie", or early steam drifter, heads between them while the crew on her foredeck lower her mast which is used only in port as a crane for unloading fish.*

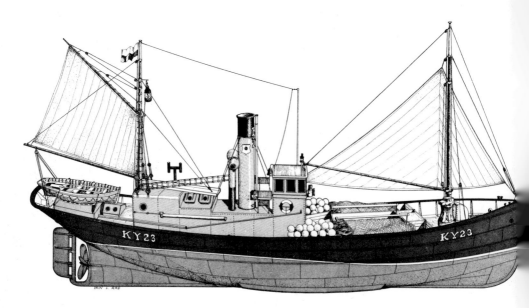

Silverdykes
Scottish steam drifter, c.1910

swivelling derrick, with which the fish were swung up to the quay in quarter-cran baskets.

The small galley was aft of the casing and provided access by ladder to the engine room and to the small but comfortably fitted-out stern cabin where the ten crew slept in 'cupboard' bunks and ate at a triangular table. The men lived well, being waited on by the cook, usually a 14-year-old boy who scrubbed the cabin, shopped for fresh provisions and prepared the dinner while the catch was being unloaded in the morning – every hour making a robust brew of tea delivered to the men in tin mugs.

The evening meal, a ten-pound leg of mutton roasted in the oven with boiled or roasted vegetables, followed by duff and treacle or jam, was dished up as the drifter steamed thirty or forty miles out to the fishing grounds, and the crew broke from their work preparing the nets to eat in pairs. After shooting the nets, and scrubbing out the fish hold, the men had a supper of cold left-overs and bread and cheese and tea. They then turned in for a few hours' sleep before the hard

72. Almost awash in a stormy sea, one of the first Lowestoft steam drifters built in the 1890s runs homeward between the pier-heads.
73. Steam rises from hot cutch as nets are "barked" to preserve them aboard steam trawlers at Peterhead.

74. *Steam drifters race for Great Yarmouth, their holds full of herrings, c.1924. In these circumstances safety valves of boilers were usually weighted down for extra speed, often causing them to blow up.*

work of hauling commenced in the early hours, leaving one man on watch. Ten herrings a man were fried for breakfast, which was not eaten until all the nets were on board.

Steam power came to English and Scottish herring fleets a decade after it had swept through trawlers, but when it did come the conversion was even more rapid. In 1898 there was only one steam drifter at Lowestoft, named *Consolation* (see picture on page 246) because it was fully rigged in case of engine failure. Within two years the building of old-style 'luggers' stopped. By 1903 no less than 101 steam drifters were registered at Lowestoft; in 1909 there were 139 steamers at Yarmouth and 243 at Lowestoft.

In Scotland experiments had been made with auxiliary engines in ketch-rigged fifies since 1871 but steam power did not take over until the turn of the century. Although Scotland's conversion to steam echoed that of Lowestoft, with about 100 steam drifters in 1902 and more than 800 by 1912, the old zulus and fifies managed to hold on because they were so much cheaper to run. There was still about 2,000 sailing drifters in the east coast ports of Scotland in 1905, and 1,300 in 1910, though many were fitted with auxiliary hot-bulb paraffin engines to make them more competitive.

In the great 1913 season at East Anglia 1,163 Scottish boats took part and took thirty-four per cent of the catch. Of these, the 854 steamers earned an average £794, the 100 motor boats earned £365, and 209 sailing drifters £235. During this period it was reckoned that every steam drifter gave employment to 100 people, counting crews, gutters, dock workers, tug crews, railwaymen, shipwrights, engineers, salesmen, curers, fishmongers, coal-miners, coopers, rope-makers, net-makers, and others. Crews were paid in a variety of ways, most taking shares in the season's profits. In Lowestoft the profit was divided into eighty-five 'doles', each one being counted one-eighth of a share. The skipper took fourteen doles (one and three-quarter shares), the mate eleven, engineer (or 'driver') ten, stoker, hawseman, net-stower and waleman seven each, the two 'younkers' (general hands) six each, the cook/boy two, and eight doles (one share) was set aside for engine maintenance.

Between the shallow banks off the Norfolk coast every drifter shot about two miles of nets until they lay in close parallel rows, with often only a few yards between them. If the wind changed suddenly every vessel drifted to leeward over the nets of its neighbour and hundreds of nets could become tangled or lost. If a gale sprang up they were driven

---

75. *A steam drifter, typical of hundreds built before the First World War, runs from Lowestoft and is caught by a curling sea.*
76. *Landing herrings from steam drifters at North Shields in the 1920s.*

down-wind, towing their nets which twisted round the warp. As the nets were laboriously hauled in, and their drag was reduced, the drifter went even faster. The crew of Smith's Knoll lightship off Great Yarmouth recovered so many herrings from nets that wrapped around the anchor chain that they had a small smoking plant on board.

There were certain signs which indicated the 'swim' of herrings when the fish congregated in shoals of billions. A drifter skipper looked for gannets, whales and porpoises, and so-called 'greasy' water when the surface of the sea was covered by a scum of oily droplets. Night after night a drifter might catch only a thousand or two herrings, but when the swim came his nets could fill with more than 100,000 fish in less than an hour. The swim was likely to be biggest near the full moon, when the shoals seemed to be stirred up by stronger tides and were attracted nearer to the surface by bright moonlight; the largest swims occurred immediately after a south-west gale.

Hauling nets on the rolling deck required teamwork and skill, and involved the entire crew except the skipper and engineer: two men pulling in the cork line and foot-rope, four around the hold hauling in the net and shaking out the herring which fell into the hold and over the side decks, the cook coiling the warp, and one man regulating the speed of the capstan and casting off the net seizings as they came up on the warp. Winning the net a yard at a time and shaking or 'scudding' was very hard work: some Scottish crews did not scud the nets until they reached harbour.

The excitement of the big herring swim, with the fleets of pipe stalkies racing twenty or thirty abreast for the narrow harbour entrance of an East Anglia port, is well described by fisheries scientist Michael Graham who was on board a drifter when it had its best shot of the season: 'It was soon clear that we had a great catch. In the second quarter of the fleet of nets there was a very great quantity of herring, all alive. One net alone had three crans in it; later there was another as good; then one with five crans; and another was good.

'By ten o'clock in the morning we were the centre of a riot of gulls, and our fellow fishermen were steaming up to look at us. There were herring scales in everyone's hair, in their eye sockets; they seemed, cemented on our fingernails; they were plastered on the funnel; they flew through the air like sawdust. Tea was brought every hour.

'One or two neighbours had herring, but not like ours. We saw them leave for port, where they would spread the news. At noon we came fast; the nets would come no more. We "gave" them, twenty of them, to a neighbour who had been waiting for this offer. He would try to lift them from the other end, and the fish would be his and the gear ours if he salved any. The herring had died of drowning because they could not work their gills, and their weight had taken the nets down to catch on something on the sea-bed.'

The arrival of drifters down by the head under the weight of herring transformed the quiet docks into a scene of chaotic bustle. With shrieking whistles and belching funnels the drifters lined the quays three and four deep and the roadways were covered with baskets and barrels brought by lorries and horse-drawn carts. At Lowestoft the saleroom – known as the hippodrome – was at the end of the market; samples were spread out on trays beneath the salesmen's desks and buyers foregathered in throngs waiting for the first salesman's bell to signal the beginning of the sale at nine o'clock.

The first shots were snapped up by Germans who could afford to pay a little more because the boats were next to the quay and with steamers waiting on time charter speed was essential. Their herrings were placed in oblong white-wood boxes, sprinkled with ice and salt, then nailed down and taken by cart to the steamers which sailed as soon as the tide served.

As prices fell the curers had their opportunity to buy thousands of crans for pickling. Meanwhile drifters kept arriving and the pressure on quay space and water area became greater, with boats overflowing into the trawl basin or the yacht harbour. On the pickling plots the Scots lassies 'gipped' the fish, singing as they dipped to the big troughs with a rhythmic swing. Flaring paraffin 'ducks' or naphtha lamps blazed out as the short autumn day merged into twilight, and in their flickering glare fishermen in scaly, shining oilskins and sou'westers, working amid clouds of steam, had a strange and fantastic appearance. Then, as the activity gradually died down, the work complete, the pungent smoke of oak billets in the curing houses skirled through the still air.

Dover Patrol.

When war came in August 1914 the Royal Navy's first impulse was to despatch steamers to the fishing grounds to recall all fishing boats, and to clear the seas for battle. Destroyers could be crippled by drifter nets, minefields could be fouled and thus revealed to the enemy, cruiser squadrons meeting a scattered fleet of trawlers with their gear down might have to turn away at a critical moment of pursuit, and the presence of drifters and trawlers would provide the enemy with opportunities for disguise.

Fishermen did not flock to the recruiting stations. The Admiralty had little patience for the fisherman's traditional independence and dislike of regimentation, and there was not a trawler skipper afloat who would admit that any naval officer was a real seaman. In 1911 the Admiralty had negotiated with owners to form a reserve of trawlers for use as minesweepers in the event of war, but when a training scheme for a reserve of manpower was started only eleven fishermen enrolled and one of these was rejected for defective eyesight. But when fishermen saw that they were needed they joined up readily, forming crews for 1,467 trawlers and 1,502 drifters that were requisitioned and flew the White Ensign as minesweepers and patrol boats. Hull and Grimsby alone supplied nine thousand men and eight hundred trawlers. Ultimately, fishermen made a greater contribution to the war effort than any other community. Miners contributed thirty-six per cent of their work force, mainly to the army, but forty-nine per cent of fishermen served in the navy while many others joined the army.

Skippers received warranted commissions and deckhands were rated as able seamen, but their loathing of regimentation never altered and 'regulars' were at first outraged at the sight of double-breasted naval uniform jackets being worn over fishermen's jumpers, and so-called officers drinking with mere bosuns and cooks. Then there was a crisis over saluting, and a feeling of ill-will persisted on both sides until everybody became so busy that there was no time to worry about it.

Of the 14,000 men who remained in fishing throughout the war, running the same risks at sea as any man on war service, more than 8,000 were too old to fight and about 400 were too young; of the remainder, about a quarter were unfit for military service. At first they were contemptuous of Admiralty attempts to control them. Their attitude was neatly summed up by the government report on sea fisheries for the war years: 'As a rule you could not tell a fisherman he risked violent death; his instinct and training bade him follow the fish whatever the risk; the risk he was not prepared to take was that of standing idle on the quay'.

Fishing in the North Sea was severely curtailed. The immense amount of plant at Hull and Grimsby had to lie idle or be adapted for war-time uses. Fleetwood and Milford Haven became the principal trawler ports. After its record year in 1913 the East Anglia herring fishery stopped dead, not just because of the risk to vessels from mines, submarines and zeppelins, but because prices slumped since eighty-six per cent of the catch would in normal times have been exported to Baltic countries, many of which were now enemies. The government issued leaflets to tell householders how to cook salted herring in the hopes that consumption would rise.

Landings of all fish fell from 1.2 million tons in 1913 to 907,000 in 1914, and only 390,000 tons in 1917. By the end of the war, when catches began to increase a little, the average price of fish (to the fisherman) had risen four and a half times, from 11s. 1d. in 1913 to 48s. 11d. in 1918. Cod went up by eight times, herring by seven. But the risks of fishing were high: in the first three months of the war Grimsby alone lost twenty vessels by mines, and in the first year lost seventy-three vessels.

A total 675 fishing boats and 439 lives were lost during the war. In 1917 vessels were captured or sunk at the rate of about four a week, among them were 156 trawlers including a whole group of eight on their way to Iceland. However, by then guns were becoming available and one or two trawlers in each group were armed. One trawler was also fitted with wireless so a group could be on the receiving end of naval intelligence instead of just reporting ashore by releasing carrier pigeons. Once the scheme got going its success was complete and despite two encounters, in each of which the trawlers had the best of it after some desperate fighting, only four trawlers were sunk by submarines in 1918.

With no weapons and no escorts for most of the war, trawlers could rely only on resource and cunning. One that got away from a submarine was the Grimsby trawler *Pinewold* in which the crew made a show of abandoning ship after being stopped by a submarine but secretly stoked up the boiler to build up a head of steam, then made their escape. But when the trawler *Plymouth* tried exactly the same trick soon afterwards the submarine put a shot across her bows and the crew had five minutes to pull clear before the trawler was sunk by shell-fire.

The diminishing fleets of sailing smacks suffered cruel fates. 127 smacks were sunk in the North Sea, 62 in the Channel and 60 on the west coast, a total of 249, with a loss of fifty-three lives. U-boats did not waste torpedoes but surfaced nearby and sent over a small boat with armed men who laid time charges. The crews were cast adrift in their boats. About 178 smacks were scuttled in this way, 54 by gunfire, 8 by mines, and the remainder by other means such as fire. In the third week of March 1917, 21 fishing boats were sunk, and in one night a single submarine blew up eleven small herring boats in the Firth of Forth, one every sixteen minutes, leaving one boat afloat for the crews. – 'a policy of deliberate murder of inoffensive fishermen and destruction of their means of livelihood.'

Before sinking trawlers German sailors sometimes raided the pantry, took baskets of fish, or unscrewed brass fittings. When the smack *Boy Jack* was stopped by a U-boat she was ransacked for iron, copper, rope, side-lights, two baskets of soles, four turbots, a conger eel, biscuits, flour, and tea; the boarders even cooked some of it before placing the charges. Skipper Harry Howe later told *The Times* what happened next: 'They took us to the sub and we stood on deck, half dressed. They asked the name of the smack and questioned us about the coast, and laughed at us. Then they photographed the smack and sank her with a bomb, and deliberately smashed our little boat. When we had been standing on the deck about three hours they sank a Belgian trawler after letting us believe we were to be put aboard her. The Belgians joined us, three of them and five of us.

'All the time the Germans kept a good look-out through glasses, then they saw a British patrol boat. I heard the commander slap the conning tower. The Germans went below, leaving us on the deck. The submarine submerged quickly. We had no warning. I saw our mate on top of the water. I never saw the cook. The deckhand gave up at once: he was a good swimmer but he couldn't get his seaboots off and lost his nerve; he was as good as raving mad. I was practically unconscious when the patrol boat picked us up. Three of our men and the Belgian skipper's son were drowned.'

It didn't always go the submarine's way, however. In January 1916 the Lowestoft smack *Acacia* was fishing when a submarine fired on her with a machine-gun, then approached. Skipper J. Crooks had lost his first smack the previous August and with a fine south-west wind blowing he decided he could fight. With a few blows the trawl warp was suddenly axed through and the smack leapt forward 'like a spurred horse'. The skipper put his helm up and tried to ram the submarine but missed by only a few feet; it crash-dived and slunk away, and Crooks was later awarded fifty pounds by the Admiralty.

Early in the war the small U-boats were so brazen in their behaviour, stopping, boarding and scuttling one fishing boat after another, that four Lowestoft smacks were fitted with semi-concealed three-pounders. Within two days of the first one leaving harbour a U-boat was sunk. More decoy smacks were chartered by the Admiralty and one was credited with five sinkings and two possibles. This resulted in the submarines intensifying their attacks on smacks and no longer giving crews time to make their escape before opening fire.

It was in one of these 'Q ships' that a fishing skipper won the Victoria Cross. Skipper Thomas Crisp R.N.R., of the smack *Nelson* (formerly the *I'll Try* in which he had sunk a U-boat a few months earlier) was mortally wounded in a fierce fight with a submarine. As he lay on the deck he continued to command his little ship until the ammunition was exhausted and she started to go down. Then he ordered the mate, his son, to throw overboard the confidential papers. As the crew abandoned the smack in the small boat he refused to be moved. His last words were to his son: 'Tom, I'm done – throw me overboard.'

If the Royal Navy saved the Allies, it was undoubtedly the fishing fleets that saved the navy. Lord Jellicoe stated it categorically in February 1918: 'The Grand Fleet could not exist without trawlers.' The Germans pinned a great deal of faith on mines and laid them in such numbers that seaborne commerce ought to have been paralysed. That the sea-lanes were kept open at all was due almost entirely to trawlers which were converted into minesweepers, with their skippers and deckhands transformed into naval officers and able seamen. A total 444 vessels were lost while minesweeping, with 234 officers and 2,058 men, mostly fishermen. Of the 394 fishing vessels lost on naval service, 246 were requisitioned trawlers, 130 requisitioned drifters, and 18 Admiralty trawlers.

Of all sea-going jobs in war-time, in the Navy or out of it, minesweeping was undoubtedly the most consistently harrowing. The enemy was unseen: there was no question of exchanging shot for shot, relying on skill and strength and cunning to get you through, as in a battle. A minesweeper was a decoy for lurking mines. Day or night, every breath you drew could be your last. First Lord of the Admiralty Sir E. Carson likened the crew of a minesweeper to the soldier who daily goes over the parapet – 'He carries his life in his hands at every moment and he does it willingly.'

Owing to the shortage of destroyers, trawlers played a significant role in the Dardanelles, first attempting to sweep the straits of mines under the searchlights and heavy guns of formidable modern forts, then searching for submarine bases, and finally ferrying troops and supplies to the beaches.

It was in the Adriatic that the second fisherman's V.C. was won. The *Gowan Lea* was one of a fleet of steam drifters tending a fifty-mile blockade of anti-submarine nets in the Straits of Otranto on May 15,

1917 when an Austrian cruiser appeared nearby in the darkness and ordered the crew to stop and abandon ship. Although the great ship bristling with guns was only one hundred yards away, Skipper Joseph Watt R.N.R. ordered full speed ahead and called for three cheers and a fight to the finish. Firing her tiny gun, the 35-ton drifter charged at the cruiser. The enemy's first shot disabled the breech. Under heavy fire the crew tried to repair it and one man was injured when an ammunition box blew up. The cruiser passed on and the gallant little ship was able to rescue the survivors of the drifter *Floandi* which was next to be attacked and received a broadside. Fourteen drifters were sunk by three light cruisers which attacked the blockade that night. Five other fishing skippers received the Distinguished Service Cross.

The Dover Straits was the lifeline of British forces abroad. Thirteen million men, two million horses, half a million vehicles, twenty-five million tons of ammunition and supplies and fifty-one million tons of coal and fuel crossed the Channel from Dover, and it was fishermen in drifters flying the White Ensign who were the mainstay of what was called the Dover Patrol which succeeded in keeping this vital link open. Fishermen called it the suicide club.

The essence of their task was closing the gate to submarines that crept in during darkness to lay mines across the paths of troopships and supply carriers, and the hunter submarines that could save eight or nine days in their journey to the Western Approaches by taking a short cut through the straits. Drifters were thin-skinned, slow, and had no weapons suitable for offence or defence except rifles. Rear Admiral Sir Reginald Bacon, naval commander of the area, wrote of them: 'It was difficult to imagine a craft less suitable for fighting.' But in June 1915 the Dover Patrol had reached its maximum strength with 132 drifters and three yachts manned by more than 1,500 officers and men of whom less than a dozen had been members of the Royal Navy six months before.

The drifters shot fleets of ten nets, each net about one hundred yards long. The nets were made of thin galvanised steel wire with a mesh ten to twelve feet square and were held afloat (after scores had been lost using kapok) by many glass balls. The idea was to release the net as it wrapped round a submarine then call in destroyers with depth charges, but the nets were frequently fouled by sea-bed obstructions and often there were false alarms. The nets had to be hauled in every six hours and repositioned to drive with the next tide. It was tedious, dreary, dangerous work.

Then special mines were placed in the nets, and it was found that nets could successfully be moored on the sea-bed. The Admiralty thought that no submarines were getting through – until papers salvaged from an enemy U-boat showed that the defences were regarded as futile and submarines were, in fact, able to slip through day and night. To plug the gap mines were laid at ten-foot intervals and at varying depths across the straits and 'ladders' of anchored net mines were laid down by the mile. To prevent enemy submarines from sneaking through by night, on the surface, about eighty drifters constantly patrolled at intervals of about three hundred yards. Any U-boat was fired at and driven below the surface where there was a good chance of it hitting a mine. At night the minefields were made a glare of light with searchlights and enormous flares designed by the man who made the Crystal Palace fireworks. At least twelve, possibly nineteen, submarines were sunk.

However a surface patrol of small and ill-defended ships advertising their presence with lights invited attack and on February 15, 1918 it came in the form of German destroyers that slipped through British patrols and ran down the line of drifters, firing, ramming and sinking where they could. Seven drifters and two trawlers were sunk, seventy men were killed and hundreds wounded. But the following night every single drifter that should have been resting went out and the patrol was more brilliantly lit and more defiant than ever.

Throughout the First World War naval commanders were often exasperated or vexed by the casual attitudes of the fishermen who served them but they also gained an undying respect for their seamanship and their loyalty. The ambivalence of the Royal Navy's love/hate of fishermen volunteers is reflected in these words by Admiral Bacon: 'One of the greatest difficulties when dealing with these gallant fisherfolk was to impress on them thoroughly the foolishness of taking their vessels to the assistance of a ship that had been mined, since one mine invariably meant the presence of several others nearby. My orders were to send their *boats* to the rescue of the crews. The fellowship of the sea, however, often led to gallant but unwise actions. Of course, from the admiral's point of view, no captain has the right, out of mere gallantry, to risk his vessel unduly. He may risk himself: this is his own lookout. You can inveigh about the loss to the country, and admonish, but at the same time it is impossible for any sea officer, at the bottom of his heart, to condemn such deeds.'

World War to Cod War.

Since 1939 British fishermen have fought a series of wars, some on
several fronts at once. Only one was a shooting war, but all have been
fights for survival. The enemy has been U-boats and dive-bombers,
Icelandic coastguard gunboats, declining catches, rocketing costs, unfair
industrial fishing methods used by short-sighted nations, international
agreements painstakingly arrived at and openly flouted. All the main
economic stocks of white fish have been fished out or are dangerously
over-fished. Herrings have disappeared completely from the North
Sea. Distant-water trawling has become so severely depressed that
throughout the 1960s and at intervals since then it has been sub-
sidised by the tax-payer.

In the Second World War fishermen again gave massive support to
the Royal Navy. Trawlers became minesweepers and escorts, fish holds
were converted into mess decks, and the Royal Navy Patrol Service,
to which most fishermen belonged, was known as 'Harry Tate's Navy'
after the comedian who was notoriously confounded by gadgets. In
all, 816 trawlers were requisitioned at one time or another and 146
were lost on war service.

Fish became more important than ever during war-time and remained
unrationed, although prices were controlled from 1941. With the best
vessels taken up and the best fishing grounds mined, landings of fish
dropped by seventy per cent. Fleets of Dutch, Belgian, French and
some Polish fishing boats that escaped the Germans were settled at
Plymouth, Brixham, Newlyn, Milford Haven and Fleetwood; landings
by these vessels brought Britain's total fish supply up to about half
the pre-war level. Hull closed down completely as a fishing port and
Fleetwood took over from it in importance.

Drifters were at first given air cover while fishing off the east coast –
RAF pilots called it the 'kipper patrol' – but after 1939 herring fishing
practically ceased. Of 277 steam drifters more than 200 were taken
over for patrol duties, target towing, and tending barrages.

Armed trawlers escorted hundreds of small boats, many of them fishing boats of various kinds, that helped in the evacuation of 335,000 Allied troops from the beach at Dunkirk. Eight trawlers were lost. The Essex bawleys, with shallow draft, proved ideal for picking men from the beaches and six of them saved a thousand soldiers, though one was blown to pieces by a bomb.

Six years of war again had a remarkable effect on stocks of fish in the North Sea: the density of fish trebled and landings after the war increased five-fold. When peace came the shortage of trawlers and surplus of fish resulted in a swift building programme. By 1951 at least ninety new trawlers had been built for Hull and Grimsby, and fish stocks were again at the critically low pre-war level.

Between the wars great advances had been made in the knowledge of fish and it was established beyond doubt that closure of certain nursery grounds would be beneficial. Dr Michael Graham, director of fisheries research in Britain, was able to prove to I.C.E.S. that if fishing effort in the North Sea was reduced the yield would remain the same because fish would grow bigger. Agreements were signed but war intervened and in the post-war years the nations of Europe were too hungry to consider limiting fish supplies. It was not until 1954 that the foundation of modern fish conservation was laid with international agreement on net sizes. The size of mesh chosen was 75 mm. (2.9 ins.) from knot to knot, but 110 mm. (4.3 ins.) in Arctic and Icelandic waters where the main catch was cod and haddock. With modifications, the agreement has been in operation ever since, but enforcement has been up to individual countries, and some authorities, notably Britain, are stricter than others.

Pressure mounted on stocks of fish in waters covered by I.C.E.S. which in 1959 became the North East Atlantic Fisheries Convention (N.E.A.F.C.). The yield had increased from three million tons in 1913 to well over four million in 1938. The rapidly expanding Russian catches did not appear in the statistics until 1955, when the total was eight million tons. With improved methods of catching herring in the mid-1960s the total became ten million, but the total catch of cod halved in ten years. International goodwill and co-operation had been unable to save the North Sea herring, and it seemed that other types of fish would suffer a similar fate.

The story of the herring is an object lesson in the kind of over-kill that threatens all commercial fishing in the north Atlantic and the Arctic, as well as in many other parts of the world, and which in 1976 was threatening the west country mackerel.

One of the methods of herring fishing developed mainly after the Second World War was ring-netting, for which small motor boats worked in pairs. A herring shoal was completely surrounded by a large net held fast by one vessel while the other paid it out in a circle, then both crews worked on the same vessel to haul it in. At first there was great opposition to ring-netting and ring-net boats trying to land their fish at Stornoway had their baskets of fish thrown back aboard and their lines cut. When the post-war development of the echo sounder allowed herring shoals to be 'spotted' the advantages of the method could not be ignored.

From the ring net developed the purse seine, an immense net up to ninety fathoms (540 ft.) deep and 300 fathoms (1,800 ft.) long. It was shot around a herring shoal by one vessel then the two ends were drawn together forming a cylinder round the fish. Also, the bottom of the net was closed, so the fish were completely shut in. Sometimes a flashing light was lowered to prevent them escaping as the ends were drawn together. The net was then hauled in by a hydraulically driven 'power block', herding the fish into one end where they were scooped out.

Pioneered in Iceland and taken up on a large scale by Norway in 1963, it was used extensively in the North Sea in the mid-1960s. Compared to a 'large' drift-net catch of about thirty tons, an average haul for a purse net – which was big enough to enclose St Paul's Cathedral – was 200 tons. In two years the total European herring catch doubled. Soon Norway had a fleet of five hundred purse-net vessels: they made a killing near Shetland in 1965 when a small fleet took in ten days more than the British herring fleet in ten weeks.

Most of the fish caught by this method was converted into fish meal to feed livestock. Its disadvantage was that all fish, big and small, were scooped up. The result can be imagined. By 1969 Norway's herring catch had dropped from 600,000 tons to less than 80,000 tons. The Danes, fishing herring nursery areas near Jutland, saw their catches drop from 1,000,000 tons to 22,000.

The East Anglia herring season, which had been declining since the 1920s, was already in a bad state. In 1959 only 106 boats fished from Great Yarmouth and Lowestoft and the whole season's catch of 50,000 crans had in the past been exceeded in a single day. In 1962 there were only ten Scots fisher-girls at Yarmouth, and in 1968 only five Scots drifters bothered to come. The following year, probably for the first time for many centuries (discounting war-time) there were no herring drifters at East Anglia.

---

77. *Almost hidden by an Arctic swell, a Hull trawler belches smoke as she races for the Humber.*
78. *St Andrews Dock, Hull—after unloading her catch of 2,000 ten-stone kits of prime fish the trawler* Kingston Onyx *is moved by tug to a berth where she will be bunkered for her next trip north. This vessel is typical of about 240 side trawlers that comprised Britain's distant-water fishing fleet until the advent of stern trawlers in the mid-1960s; many are still in commission.*

By 1976 herring no longer existed in commercial quantities in the North Sea apart from a small area inside British territorial waters off North Shields. Stocks in the Irish Sea were falling alarmingly low and conservation measures were observed only by British vessels. The only herring fishery left in European waters was in The Minch, in British waters, between the Outer Hebrides and north-west Scotland. However the spawning grounds of these herrings lies west of the Hebrides, in international waters. These are protected by international agreements on quotas that may be taken by member countries, and Norway and Denmark have agreed to fish only for human consumption. In 1974 British herring boats landed 150,000 tons, mainly from The Minch; Lowestoft's share was barely one per cent, Yarmouth's nil.

The kind of fishing boat used to catch herrings and fish in inshore waters since the 1930s is what we know today as the 'typical' Scottish M.F.V. (motor fishing vessel). It was the result of a kind of marriage between the iron-built steam drifters and the cheapness, handiness and adaptability of the craftsman-built motorised zulus and fifies. Built of wood with a high bow, cruiser stern, and a large wheelhouse, and just under eighty feet long to fit the official definition of an 'inshore' fishing boat, this seaworthy and powerful type of vessel could handle a variety of different types of fishing gear – seine nets, trawl, long-lines, drift-nets, ring-net, or (in the case of modern boats) purse seine.

As fishing became less economic in the 1930s, a new type of net was introduced which exploited the inshore areas where the seabed was unsuitable for trawling. This was the Danish seine, shaped like a small trawl, each warp of which is nearly a mile long. The boat shoots first one warp, then the net, then the second warp, returning to its starting point with the net and warp laid out on the seabed in a pear shape. Then anchoring or stemming the tide by engine, it gradually winches in the warps so they come together, 'tickling' the fish towards the centre. When the warps become parallel the net begins to advance and scoops them up. Usually done in daylight and always in calm or moderate weather, each haul takes about three hours and requires much patience and perseverance, but on the Dogger Bank a skipper could expect steady catches of eight to ten stones of best plaice every haul.

This type of fishing proved ideal for a large number of small motor vessels and steam drifters which found they could catch as much white fish in the North Sea as a trawler. Moreover, unlike trawlers, they were not restricted from fishing within the three mile limit.

Fishing the whole of the North Sea as far as the Norwegian coast, and ranging along the entire west coast of the British Isles, so-called inshore vessels were developed with the help of grants and loans into the most successful branch of the industry. Up to 1969 boats using seine nets increased their landings by more than twelve times to 100,000 tons, and the value of this catch increased 300-fold. Freezer-

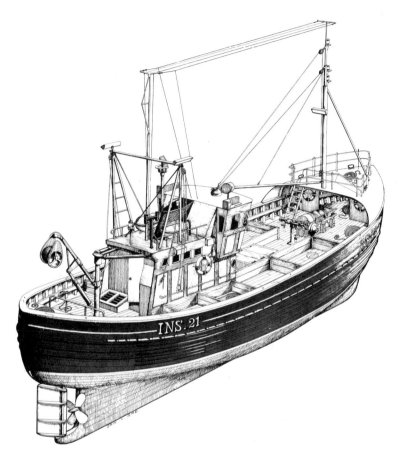

Horizon
*Lossiemouth wooden motor seiner typical of the type built
since the 1950s*

lorries allowed fishing to become even less centralised: a handful of
boats could work from any port with road access. In comparative terms
their earnings made those of the big companies seem insignificant, but
in ten years the inshore fleet's share of the catch went up from twenty-
eight to fifty-five per cent, more than that of the entire deep-water
fleet. In 1973 there were 1,832 inshore vessels (86 brand new), compared
with 488 larger trawlers and factory ships.

In the early 1950s seine-net fishermen in the Firth of Forth discovered
that a prawn-like creature commonly called Dublin Bay prawn or
Norway lobster, which previously had been discarded as valueless, in
fact commanded a ready market as scampi. As a result the trawling
ban which had been enforced in Scottish waters since 1889 was relaxed
and by 1969 the scampi catch was worth more than £2 million a year.

Although vessels were fewer and purchase costs well into six figures, the 1,000 inshore boats in Scotland have brought renewed prosperity to its fishing community – an echo of the boom years before the First World War. As in the days of sail a hard-working and talented young man can again become owner of his vessel, with a thirty per cent grant, a fifty per cent loan and the rest his own money. With more than a quarter of a million pounds at stake his own share of twenty per cent is substantial, but older boats could be bought much more cheaply. A fisherman can expect to earn at least £5,000 a year, a skipper twice as much. Close family and local links remain among the crews, and when their vessels are away at other ports the crews return home for weekends by minibus.

While the inshore fleet sailing from small ports serviced by lorries prospered, the great distant-water trawling industry in the big ports began to founder. The most significant post-war development was the freezer-trawler introduced in the mid-1960s; in 1975 there were thirty-five in Hull, six in Grimsby and one in Fleetwood. These are big vessels 300 feet long, almost ocean liners compared with the *Zodiac* which had started it all in 1881. On a huge factory deck beneath the main deck fish are gutted, headed, filleted and deep frozen in blocks, all within minutes of being caught. The ships stay at sea for as long as ten weeks, the crew having recreation rooms where they can watch films, and comfortable cabins. Their catches are normally contracted on an annual basis to frozen-food companies.

Like the freezer trawlers, the new generation of 'wet-fish' (i.e. non-freezer) trawler has a sloping ramp in the stern up which the net is hauled, greatly simplifying the handling of heavy gear. The crew are able to work in comparative shelter, protected on either side by high bulwarks and forward by the main superstructure. The modern trawler skipper has an indoor job. The ship is steered by automatic pilot, electronic fish-finders draw pictures of shoals of fish under his keel, radar has done away with the traditional fishing marks and he still does not require a sextant because he can read his latitude and longitude from the dials of the Decca navigator which is linked to radio beacons. But phlebitis is still an occupational disease.

The crew of a modern trawler is not so lucky. They might have individual cabins, basins and showers with hot and cold water, and film shows, but outside the porthole is the Arctic winter and all the dangers of ice and storm. All but a handful of Britain's 334 middle and

---

79. The new style of fishing—small "inshore" vessels (which range as far as the coast of Norway) using small ports and unloading their catches into lorries which in remote ports have refrigerated storage; two seiners (foreground) and a small trawler lie in Amble, Northumberland.

80. *Force 9 Gale, north of Norway, mid-winter—
Grimsby trawler* Vivaria *runs with the weather.*

81

82

distant-water trawlers are still the old 'side-winder' type and the men have to work on open decks. In 1969 Jeremy Tunstall wrote in *The Times*: 'Trawling is an occupation in which the working conditions have to be seen to be believed. Most of the Hull and Grimsby trawlers are still the conventional sort which shoot the net over the side of the vessel – and that means turning the ship sideways to the weather. Deckhands still work on an open deck; there is only a very low rail; winches and other machinery are virtually unguarded. A trawler bucks like a wild thing in even a stiff breeze; in a gale it dips its rails under every wave, the driving spray envelopes the bridge windows, and at every so many waves the ship lurches sickeningly – like a horse missing a heartbeat as it takes a jump. During gales the deckhands lie jammed into their narrow wooden bunks bracing their bodies against the movement. When the wind eases they are out on the deck again – and in winter the deck is coated with ice, lit by floodlights shining out of the perpetual Arctic night.'

As far as the individual fisherman is concerned the hardships and dangers of the job are as close and as personal as they ever were in the days of sailing smacks on the Dogger Bank. Ships are bigger, more comfortable to live in, but gear is heavier and more difficult to handle, and the fact that vessels can work tougher conditions means that men, too, have to be able to cope. Despite more than a century of progress the 'fisherman's walk' remains at three steps and overboard – in icy Arctic waters a man dies of shock, in seconds, before he drowns.

Sometimes all the sophisticated electronic gear of a trawler, the strength of a modern hull and the power of its engine, are not proof against the hazards of the sea. In January 1955 trawlers sheltering at Iceland heard a terrible drama unfold on their radios as the Hull trawlers *Lorella* and *Roderigo* steamed in company to aid another which had fouled her propeller. The lame trawler managed to free herself and seek safety but her two would-be rescuers found themselves in hurricane winds with freezing temperatures that persisted longer than usual. As ice accumulated on their super-structures they had to steam faster into the wind to prevent themselves from broaching. The faster they steamed the more spray was created, building more layers of ice. Visibility was practically nil, the seas mountainous.

At 2.39 p.m. the *Lorella* rolled over under the weight of ice and sank. *Roderigo's* skipper radioed at 5.05 p.m.: 'Still going over to starboard,

---

81. *Build-up of ice on the deck and superstructure can capsize a trawler if the crew is unable to clear it with axes and steam hoses.*
82. *Ashore on a rocky coast, crew awaiting the dawn and anxious for rescue—the modern Leith trawler* Summerside *has run aground despite radar and other navigational aids.*

cannot get her back.' 5.08 p.m.: 'Still going over, going over.' The words 'Roderigo going over' were repeated in morse for another four minutes then transmission ceased. Forty men were lost in the two ships.

The public conscience was stirred in 1968 when three trawlers sank in exceptional weather conditions in the space of a few days with a loss of fifty-eight lives. 'I am going. We are laying over. I am going. Give my love and the crew's love to their wives and families' – these were the last words from the *Ross Cleveland* as she listed over and over in mountainous seas under the weight of tons of ice until she capsized and sank.

In the resulting enquiry into the whole question of trawler safety some extraordinary figures and facts came to light. Mortality among fishermen was seventeen times greater than that of the average male worker, six times greater than that of coal miners. In 1963 one man in four suffered an injury. In a seventeen-year period four per cent of the work force – one man in twenty-five – died at his job. Of forty-three ships lost between 1957 and 1966, twenty-six resulted from strandings and collisions due to faulty navigation. Men had to work 180 hours in ten days on the fishing grounds in order to earn one-third more than the average shore worker and their fatigue reminded one expert of what he had seen among soldiers during the retreat from Dunkirk.

The committee formulated eighty-three recommendations ranging from the urgent need of a permanent support ship to provide meteoro- logical, medical and technical assistance during the winter off north- west Iceland (this has been provided by the government since 1968/69), the need for at least six hours of unbroken rest in twenty-four, and a recommendation that deckie-learners should not make their first trips to sea in a side trawler in winter. However the fisherman's traditionally casual attitude towards safety ('We're used to be drowned') was recognised and deplored: 'The problem is fundamentally one of chang- ing human attitudes at every level of the industry: ... Fatalistic attitudes towards accidents are a particular problem ... The majority of accidents in the fishing industry – whether to ships or to individuals – arise to a greater or lesser degree from human errors and may be prevented.'

In February 1974 the Hull trawler *Gaul*, a modern freezer trawler equipped with every navigational device and all safety gear, was lost without trace with all hands somewhere north of Norway during a bad storm. Was she pooped by a giant wave which broke over her stern and filled the ship, carrying her to the bottom? Did she tear her

***

83. *Arctic waters, mid-winter, early afternoon—aboard the Grimsby trawler* Vivaria *gutting and hauling goes on day and night while thousands of gulls feed on the offal thrown over the side.*

bottom out on a rock and go straight down? Did the Russians capture her because she was secretly equipped with electronic 'listening' equipment, as some people have speculated, or did she collide with a NATO or Russian nuclear submarine and go straight to the bottom? So far, there is no answer.

For fishermen, facing death is part of the job. The possibility that the distant waters they have fished since the turn of the century could be closed to them, or completely fished out, worries them more than risks of death or injury. In Icelandic waters, which provide half of all trawled fish, the average catch for one hundred hours of fishing dropped from 1,060 cwt in 1956 to 529 cwt in 1961. In the same period catches at Bear Island and Spitsbergen fell from 2,493 cwt to 1,421 cwt. Big fish became rarer; the majority caught were graded as small.

The situation also alarmed the people of Iceland. With no agriculture on their island of rock and volcanic ash the 220,000 Icelanders depend entirely on fish. In 1952 its fishing limit was extended from three to four miles, for its own trawlers and seiners as well as foreigners. Britain hit back by banning the landing of fish from Icelandic vessels for four years and until alternative markets were found Iceland lost a quarter of her trade.

Three miles had been tacitly accepted as a reasonable limit since the eighteenth century. It was based on the maximum range of a cannon shot and had worked well because ships could come close enough to the coast to identify navigation marks and find sufficiently shallow water in which to anchor without contravening territorial rights. The English and French met to sort out disputes in the channel in 1837, and a general fishing limit of three miles from the low-water mark was agreed. In 1881 fishing limits were accepted internationally at a meeting in the Hague and Denmark accepted three miles for Iceland, which continued after the latter became a sovereign state in 1918.

In 1958 the limit was again extended, from four to twelve miles. The issue became a test case of a state's right to expand its control seawards. Iceland claimed a special right because of her unique dependence on fishing and the urgent need to prevent over-fishing. For a year and a half British trawlers ran the gauntlet of Icelandic 'gun-boats' towing

---

84. Fisherman of the 1970s—a weather-tight bridge deck equipped with every facility including fish-finders, radar, direct engine control, radio and Decca navigator.

85. Fishing trawler of the 1970s—the Hull stern-trawler Marbella *freezes her catch at sea, the men working in shelter behind the bridge and the trawl hauled aboard by a ramp in the stern. With the new 200-mile fishing limits such large vessels have a doubtful future and in 1976 many were losing money and being laid up.*

grapnels to cut through trawl warps. When Icelanders tried to arrest and board British trawlers potatoes and flour bombs were thrown, and fire hoses used to repel boarders.

When the second United Nations Law of the Sea Conference met at Geneva to settle the vexed question of fishing rights its result was inconclusive and Britain had to give in to Iceland, though with some concessions. Half the countries at the conference wanted a twelve-mile limit and the other half backed a compromise solution of six miles, with the right to exclusive fishing of the contiguous zone extending another six miles subject to nations accustomed to fishing in that zone being entitled to do so for another ten years. Britain doubled her own fishing limit to six miles in 1964, with a further six-mile zone in which she had control over the fishing, and most other European countries followed suit.

Trouble came again in 1971 when Iceland unilaterally extended her fishing limit to fifty miles. This was more serious because ninety per cent of the catch was found inside this area. Again trawlers fished in close groups guarded by Royal Navy frigates while Icelandic coast-guards prowled round the fleet ready to cut the fishing gear of any stragglers. Previously it had been agreed with Iceland that any further dispute would be referred to the International Court of Justice at The Hague, but when Britain and West Germany presented a joint case no Icelandic representative turned up on the grounds that the court did not have jurisdiction, and Iceland's fifty-mile zone was declared illegal.

With nations everywhere pressing for greatly increased territorial rights, mainly because of oil and minerals found on continental shelves, Britain had to give in and a two-year agreement gave her the right to take 165,000 tons of fish a year from inside Iceland's limit. In 1975 the agreement ended and the limit was extended to two hundred miles, and trawlers were once more herded into boxes by frigates which literally rubbed shoulders – with much grinding and buckling of bulwarks – with Icelandic gun-boats.

Iceland's action may have been illegal, but, as before, it did not go against the grain of world opinion. The third cod war with Iceland in 1975/6 was not about the 200-mile limit, it was about the amount of fish Britain was entitled to catch in Icelandic waters before the next United Nations Law of the Sea Conference discussed the issue and

86. The cod-end is winched up the stern ramp of a modern trawler. The fish is dropped into the deck below then gutted and frozen on a production-line principle by fishermen working in shelter.
87. Modern stern trawlers are not always as comfortable as they look; here a large modern vessel is pooped by a breaking wave which ran up the stern ramp and flooded aboard.

Britain could extend her own limit to 200 miles.

By 1975 Newfoundland and Greenland waters were practically barren and in the north-east Atlantic all fish were covered by quotas. Most of Hull's large and expensive freezer trawlers spent part of the year tied up because their quotas were filled. The whole system relied on the honesty of individual nations. British fishermen regarded themselves as the honest men of Europe and respected the quotas. Other nations cooked their books. According to the British Trawlers' Federation the West Germans filleted and skinned cod at sea and landed them as haddocks, the Russians reached their quotas then carried on fishing as if quotas did not exist (who can keep a check on the Russians?), and Bulgaria, catching huge quantities, did not even come to the conference table to discuss the question.

All countries bordering the north Atlantic, including Russia, Canada and U.S.A., had extended their fishing zones to 200 miles by March 1977. The Common Market countries collectively claimed territorial rights over a zone extending 200 miles seawards from the entire coast of Greenland as well as the Atlantic island of Rockall and their own sea borders. Of this area, two-thirds was contributed by Britain. At negotiations in Brussels early in 1977 Britain was sticking out for an exclusive 50-mile zone to compensate her fishermen for loss of fishing grounds which now belong to Iceland and Norway, but the Council of Ministers was expected to assign catch quotas by species and nationality. Control of these quotas will prove as difficult as ever, and once more "foreigners", who have already fished out their own waters, will be seen sweeping their nets right up to the home beaches of British fishermen.

If they can settle their own differences, Britain and the Common Market countries could nevertheless benefit considerably from the new fishing zone. It will give exclusive rights to The Minch herrings and deep fishing grounds near Rockall, in the Atlantic. But the main problem the industry faces is to wean the public from cod and introduce new deep-water species. Cod is likely to become rare, but it is only since the turn of the century that it has been eaten in large quantities and it should not be hard to return to the mackerel, sprats, and other pelagic fish that were the mainstay of the country's fish diet only a lifetime ago.

Blue whiting alone is so prolific in Britain's 200-mile zone that it could support a fishery of one million tons a year: it is small and bony, and spoils easily, but methods of processing it already exist. A number of exploratory voyages have been made to the deep waters, fishing in six hundred fathoms instead of two hundred. Deep-sea species tend to be thin and spiny, and look like something in a horror film. They include the scabbardfish which is eaten in Portugal, can be smoked, and is said to taste like plaice; the Grenadier could be a replacement for cod though it has a lower yield and it is eaten in Russia; others have a junkety consistency but a pleasant taste.

With a different mix of fish on the fishmonger's slab and in the deep-frozen packet the fishing industry in Britain can still be prosperous – an inshore fleet doing better than before, a limited deep-water fleet fishing traditional grounds under licensing agreements, and other deep-water vessels making a living by fishing for new species.

For European nations the big problem will be the protection of their enormous fishing zones. Russia's factory-style fishing fleets were expected to net about 500,000 tons of fish in what is now the European fishing zone during 1977. Russia has never recognised the negotiating authority of Brussels and was refusing even to talk about fishing rights. The result is that Britain could find herself in the front line of yet another fishing war, this time on the defensive, as Iceland was. Royal Navy frigates will now be trying to prevent Russian vessels from fishing in her waters: this would be an ironic reversal of fortune.

While the political battles are fought on another front, the future of the British fisherman is held, as ever, in his own hands. Will he adapt quickly enough to new fishing grounds, new fishing methods, new markets, and even new fish?   In the past flexibility and a willingness to strike out in new directions has not been the fisherman's strong point: if that can change, his future in the New Europe is assured.

BRITISH FISHING PORTS

FAEROE ISLANDS

SHETLAND
Lerwick
ISLANDS

ORKNEY
ISLANDS
Kirkwall

CAPE
WRATH
HEBRIDES
LEWIS
Stornoway
Wick
HARRIS
Dunbeath
N. UIST
Helmsdale
Ullapool
BENBECULA
THE MINCH
MORAY FIRTH
Buckie
S. UIST
SKYE
Lossiemouth Banff
Fraserburgh
Peterhead
BARRA
POINT OF
ARDNAMURCHAN
Aberdeen
Findon

OUTER
Arbroath

LOCH FYNE
Anstruther
Glasgow
FIRTH OF FORTH
Edinburgh Dunbar
Eyemouth
Berwick
FARNE ISLANDS
Craster
Newcastle
Sunderland

N O R T H

S E A

ISLE OF
MAN
Hartlepool
Staithes Whitby
Robin Hood's Bay
Scarborough Filey

Dogger
Bank

MORECAMBE
BAY
Fleetwood
CELTIC SEA
Blackpool
Hull
HELIGOLAND

Holyhead Hoylake Liverpool
ANGLESEY
THE HUMBER *SPURN HEAD*
Grimsby
Caernarfon

CARDIGAN
BAY
Aberystwyth
Boston *THE WASH*
✧ *SMITH'S KNOLL*
TEXEL

Great Yarmouth
Lowestoft
Ijmuiden

Milford Haven

BRISTOL CHANNEL
Wivenhoe
Tollesbury
Brightlingsea
Barking Leigh-on-Sea
London
Greenwich *NORTH FORELAND*
Whitstable Ramsgate
Ostend
Exeter Hamble
Newquay Teignmouth Poole Emsworth Folkestone Dover
St Ives Polperro Torquay Brighton Hastings Calais
Sennen Plymouth
Cove Brixham
Newlyn Mevagissey Hallsands Boulogne
SCILLY IS. Penzance Falmouth

CHANNEL
ISLANDS

0   100

Miles

APPENDIXES

Fishermen's measurements   314

Fish landed in Great Britain by British vessels 1886–1974   315

Herrings landed in Great Britain by British vessels 1886–1974   316

Cod landed in Great Britain by British vessels 1886–1974   317

Fishing vessels in Great Britain 1872–1970   318

Fishing vessels in Great Britain in 1970 (by type)   319

Fishermen and fishing vessels in Great Britain in 1886   320

Fish consumption per head in Great Britain   320

Fish landed in Great Britain by British vessels in 1886 (by type) 321

Fish landed at selected ports in 1886 and 1974   322–3

*All figures are derived from Fisheries Statistics and Inspectors'
Reports published by H.M. Government*

FISHERMEN'S MEASUREMENTS

4 herrings	=	1 warp
33 warp	=	1 hundred (actually 132 fish)
10 hundreds	=	1 thousand (1,320 fish) = 1 cran
10 thousands	=	1 last (13,200 fish)
1 cran	=	4 baskets
	=	$37\frac{1}{2}$ imperial gallons
	=	$3\frac{1}{2}$ cwt (approx)
	=	100 'long hundreds'
	=	1,320 herrings
1 barrel	=	$26\frac{2}{3}$ imperial gallons
1 wash (i.e. of whelks)	=	21 quarts
1 kit (i.e. of haddocks)	=	2 cwt

TONS OF FISH LANDED IN GREAT BRITAIN
BY BRITISH VESSELS

Year	England and Wales	Scotland	Total	Value
1886	320,000	236,000	556,000	£5,500,000
1890	305,000	268,000	573,000	£5,900,000
1900	430,000	268,500	698,500	£8,900,000
1910	655,900	435,500	1,091,400	£11,000,000
1920	729,200	316,300	1,045,500	£27,800,000
1930	783,900	178,600	962,500	£17,200,000
1938	776,800	274,900	1,051,700	£16,200,000
1950	629,900	252,800	882,700	£35,700,000
1960	536,200	259,500	795,700	£50,900,000
1970	535,200	412,800	948,000	£75,500,000
1974	469,700	469,500	939,200	£151,900,000

TONS OF HERRINGS LANDED IN GREAT BRITAIN BY BRITISH VESSELS

Year	England and Wales	Scotland	Total
1886	98,700	155,150	253,850
1890	61,600	175,700	237,300
1900	121,250	176,000	297,250
1910	191,400	284,350	475,750
1920	231,800	157,750	389,550
1930	199,850	170,400	370,250
1938	129,350	140,050	269,400
1950	76,450	99,850	176,300
1960	16,600	88,100	104,700
1970	14,300	124,700	139,000
1974	10,900	127,400	138,300

TONS OF COD LANDED IN GREAT BRITAIN
BY BRITISH VESSELS

Year	Total	(caught in Icelandic waters)
1886	29,600	
1890	40,250	
1900	51,150	
1910	49,000	37,250
1920	125,300	
1930	276,500	
1938	349,700	
1950	321,950	88,150
1960	309,950	
1970	338,700	107,000
1974	262,450*	96,400

** includes 49,300 tons frozen at sea; also, foreign
vessels landed 21,550 tons in Britain*

FISHING VESSELS IN GREAT BRITAIN

Year	England and Wales	(First class)	Scotland	(First class)	Total
1872	Fishing vessels of fifteen tons burthen or more in British Isles				41,723
1900	7,190	(3,176)			
1920	9,235	(3,071)	8,177		17,412
1930	7,269	(2,453)	5,969	(1,179)	13,238
1938	6,614	(1,963)	5,217	(1,764)	11,831
1950	7,192	(1,328)	5,222	(1,617)	12,414
1970*	1,063		1,079		2,142

* see next table

The definition of 'First Class' varied with the development of fishing craft over the years. In every case the definition prevailing at the time is the standard used.

FISHING VESSELS IN GREAT BRITAIN IN 1970

Vessels	England and Wales	Scotland	Total
Trawlers			
40 to 79.9 ft	377	345	722
80 to 109 9 ft	120	47	167
110 to 139.9 ft	105	68	173
140 ft + Freezers	36	—	36
Others	123	2	125
Drifter/trawlers (80 to 109.9 ft)	4	—	4
Drifters (40 to 109.9 ft)	6	14	20
Seiners (40 to 109.9 ft)	172	431	603
Purse seiners (40 to 109.9 ft)	—	15	15
Liners (40 to 109.9 ft)	56	17	73
Ring-netters (40 to 109.9 ft)	1	57	58
Other vessels (40 to 109.9 ft)	63	83	146
All vessels over 40 ft	1,063	1,079	2,142

FISHERMEN AND FISHING VESSELS IN GREAT BRITAIN IN 1886

| | England and Wales | | | | | |
	East coast	South coast	West coast	Total	Scotland	Total
Fishing boats	5,458	2,567	801	8,826	14,391	23,217
Men and boys constantly employed	21,832	9,382	2,208	33,422	30,023	63,445
Part-time fishermen	5,842	5,393	1,074	13,309	21,695	35,004
Value of fish landed						
Per boat	£615	£179	£176	£448	£125	
Per fulltime man	£154	£49	£64	£118	£60	
Per all fishermen	£121	£31	£33	£85	£35	

FISH CONSUMPTION PER HEAD IN GREAT BRITAIN

Pounds per head per year (fresh, frozen or cured fish)

Pre-World War Two	21.8 lbs	1960	16.1 lbs
1948	28.7 lbs	1970	16.2 lbs
1951	22.0 lbs	1974	14.0 lbs
1955	19.0 lbs		

TONS OF FISH LANDED IN GREAT BRITAIN BY BRITISH FISHING VESSELS IN 1886

| Fish | England and Wales | | | | | |
	East coast	South coast	West coast	Total	Scotland	Total
Prime fish (soles, turbots etc)	25,150	900	350	26,400		26,400
Cod	11,750	150	500	12,400	17,200	29,600
Haddock	62,050	50	60	62,160	33,550	95,710
Mackerel	1,850	8,450	2,950	13,250	145	13,395
Herring	90,000	5,500	3,200	98,700	155,155	253,855
Pilchards		17,000	650	17,650		17,650
Sprats	6,450	950	1	7,401	1,100	8,501

TONS OF FISH LANDED AT SELECTED PORTS
IN ENGLAND AND WALES

Port	1886: Cwt	Value	1974: Cwt	Value
Berwick	58,000	£34,000		
Craster	16,000	£2,800		
North Shields	289,000	£93,000	720,000	£4,247,000
Sunderland	32,000	£19,000		
Hartlepool	54,000	£30,000	29,000	£349,000
Staithes	7,000	£3,700		
Whitby	56,000	£21,500	53,000	£626,000
Robin Hood's Bay	246	£283		
Scarborough	195,000	£85,000	83,000	£985,000
Filey	6,000	£3,200		
Flamborough Head	7,000	£4,600		
Bridlington	9,000	£3,100	47,000	£612,000
Hull	401,000	£358,300	2,823,000	£28,677,000
Grimsby	1,364,000	£936,000	2,672,000	£26,575,000
Boston	37,000	£15,000		
Great Yarmouth	737,000	£224,600		
Lowestoft	502,000	£280,000	458,000	£7,034,000
Southwold	19,000	£11,300		
Brightlingsea	30,000	£1,500		
Leigh-on-Sea	10,000	£9,100		
London	1,304,000	£950,700		
Ramsgate	33,000	£50,500		
Folkestone	38,000	£29,200		
Rye	13,000	£3,700		
Hastings	47,000	£30,300		
Brighton	12,600	£11,100		
Selsey	286	£307		
Poole	960	£1,575		
Beer	5,400	£2,400		
Exmouth	17,000	£5,700		
Brixham	48,000	£43,000	53,000	£119,000
Plymouth	142,000	£84,000		
Looe	18,000	£5,000		
Polperro	9,000	£2,600		
Fowey	2,000	£440		

Port	1886: Cwt	Value	1974: Cwt	Value
Mevagissey	255,000	£211,000		
Falmouth	26,000	£16,800		
Prussia Cove	305	£127		
Penzance/Newlyn	83,000	£40,800	50,000	£205,000
Mousehole	7,300	£2,200		
Sennen Cove	3,600	£2,000		
St Ives	102,000	£48,000		
Clovelly	11,200	£4,000		
Tenby	10,000	£7,000		
Holyhead	1,200	£1,000		
Bangor	1,400	£1,400		
Hoylake	13,000	£30,000		
Liverpool	3,200	£2,300		
Fleetwood	14,000	£9,000	1,911,000	£8,983,000
Peel (Isle of Man)	31,000	£10,000		

BIBLIOGRAPHY

The author acknowledges with gratitude his debt to a great number of books and papers which were consulted in the course of his research at the British Library, the Fishmongers' Hall, the Ministry of Agriculture and Fisheries, the National Maritime Museum, the Scottish Fisheries Museum, the National Museum of Antiquities, and at local libraries. Many Government reports and papers were very useful and are too numerous to list. Among newspapers and magazines consulted were back numbers of *The Times, Grimsby Evening News, Hull Daily Mail, Eastern Daily Press, Yarmouth Mercury, Mariners Mirror, Toilers of the Deep, East Coast Mariner, Fish Trades Gazette, Fishing News, Commercial Fishing*, and many others.

Aflalo, F. G.,	*The Sea-Fishing Industry of England and Wales*, London, 1904.
Alexander, L. Mc.,	*Offshore Geography of North West Europe*, Chicago, 1963.
Alward, G. L.,	*The Development of the British Fisheries*, Grimsby, 1923.
Anson, Peter,	*Fishing Boats and Fishing Folk on the East Coast of Scotland*, London, 1930.
	Fishermen and Fishing Ways, London, 1932.
	The Sea Fisheries of Scotland – Are they doomed?, Edinburgh and London, 1939.
	Scots Fisherfolk, Banff, 1950.
	Fisher Folklore, London, 1965.
Atkins, John,	*The Distribution of Fish*, London, 1941.
Bacon, Admiral Sir R.,	*The Concise Story of the Dover Patrol*, London, 1932.
Bagshawe, J. R.,	*The Wooden Ships of Whitby*, Whitby, 1933.
Balls, Ronald,	*Fish on the Spot Line*, London, 1948.
Barker, McKenzie, and Yudkin,	*Our Changing Fare*, London, 1966.
Bassett, F. S.,	*Legends and Superstitions of the Sea*, Chicago and New York, 1885.
Bertram, J. G.,	*The Harvest of the Sea*, London, 1865.
Boswell, David,	*Loss List of Grimsby Vessels*, Grimsby, 1969
	Sea Fishing Apprentices of Grimsby, Grimsby, 1975.
Borgstrom G. A. and Heighway, A. J.,	*Atlantic Ocean Fisheries*, London, 1961.

Brandt, Andres von,   *Fish Catching Methods of the World*, London, 1964.

Buchanan, Rev J. L.,   *A General View of the Fishery of Great Britain*, London, 1794.

Buckland and Walpole,   *Report on the Sea Fisheries of England and Wales*, London, 1879.

Burnett, John,   *Plenty and Want*, London, 1966.

Carr, Frank,   *Vanishing Craft*, London, 1934.

Carter, George,   *The Smacksmen*, London, 1947.

Chope, R. P.,   *Devonshire Fish Names*, Torquay, 1935.

Clark, G. S.,   *The Location and Development of the Hull Fish Industry*, Hull, 1937.

Clark, Roy,   *The Longshoremen*, Newton Abbot, 1974.

Close, A.,   *Fishermen's Pilot Round the British Isles*, London, 1909.

Cobb, David,   *Three Mile Limit*, London, 1950.

Collard, A. O.,   *The Oyster and Dredgers of Whitstable*, London, 1902.

Coull, James R.,   *Fisheries of Europe*, London, 1972.

Cowan, Frank,   *A Dictionary of Proverbs Relating to the Sea*, Greensburgh, P.A., 1894.

Cutting, C. L.,   *The General Principles of Smoke-Curing of Fish*, London, 1951.
     Fish Saving, London, 1955.

Dade, Ernest,   *Sail and Oar*, London, 1933.

Davis, F. M.,   *An Account of the Fishing Gear of England and Wales*, London, 1923.

De Caux, J. W.,   *The Herring and the Herring Fishery*, London and Norwich, 1881.

Defoe, Daniel,   *A Tour Through the Whole Island of Great Britain 1724–6*, edited by Pat Rogers, Harmondsworth, 1971.

Dickens, Charles,   *David Copperfield*, London, 1858.

Dickinson and Pryde,   *A New History of Scotland*, 2 vols., London, 1962.

Drummond, Sir J. C. and Wilbraham, Anne,   *The Englishman's Food*, London, 1957.

Duncan, Norman,   *Dr. Grenfell's Parish*, London and New York, 1905.

Ecclestone, A. W. and Ecclestone, J. L.   *The Rise of Great Yarmouth*, Great Yarmouth, 1959.

Edwards, George S. F.,   *The Origin and Growth of British Fisheries and Deep Sea Fishing*, London, 1908.

Elder, J. R.,   *The Royal Fishery Companies of the Seventeenth Century*, Glasgow, 1912.

Eunson, Jerry,   *A Pioneer of the Fishing Industry*, Kirkwall, 1959.

Evans, Frank,   *The Early History of Seine Netting at North Shields*, Durham, 1963.

Ford, E.,   *The Nation's Sea-Fish Supply*, London, 1936.

Fishmongers' Company,   *Report on Oysters and Other Shell Fish*, London, 1909.

Fulton, T. W.,   *The Sovereignty of the Sea with Special Reference to the Rights of Fishing*, Edinburgh and London, 1911.

Gentleman, Tobias,   *England's Way to Win Wealth and to Employ Ships and Mariners*, London, 1614.

Gibbs, W. E.,   *The Fishing Industry*, London, 1922.

Goodlad, C. A.,   *Shetland Fishing Saga*, 'The Shetland Times', 1971.

Gordon, A.,   *What Cheer O?*, London, 1890.

Graham, Michael,   *The Fish Gate*, London, 1949.
           Sea Fisheries: Their Investigation in the United Kingdom, London, 1956.

Green, Charles,   *Herring Nets and Beetsters*, London, 1969.

Green, Neal,   *Fisheries of the North Sea*, London, 1918.

Grenfell, Wilfred,   *The Story of a Labrador Doctor*, London, 1920.

Hamilton, J. L.,   *Report Upon the Fish Markets, Fish Trade Abuses and Fish Supply of the Metropolis*, London, 1890.

Hardy, Sir Alister C.,   *Seafood Ships*, London, 1947.
           Fish and Fisheries, London, 1959.
           Great Waters, London, 1967.

Hedges, A. A. C.,   *Yarmouth is an Antient Town*, Great Yarmouth, 1959.
           East Coast Shipping, Great Yarmouth, 1974.

Henderson, David S.,   *Fishing for the Whale*, Dundee, 1972.

Herbert, David,   *Fish and Fisheries – a selection of the prize essays of the International Fisheries Exhibition at Edinburgh*, Edinburgh, 1883.

Herubel, Marcel,   *Sea Fisheries – Their Treasures and Toilers*, London, 1912.

Hickling, C. F.,   *The Hake and the Hake Fishery*, London, 1935.
           The Recovery of a Deep Sea Fishery, London, 1946.

Hodgson, W. C.,   *The Herring and Its Fishery*, London, 1957.

Holdsworth, E. W. H.,   *Deep Sea Fishing and Fishing Boats*, London, 1874.

	Sea Fisheries of Great Britain and Ireland, London, 1883.
Holt, Ernest,	*An Examination of the Grimsby Trawl Fishery*, Grimsby, 1895.
Hore, J. P. and Jex, Edward,	*The Deterioration of Oyster and Trawl Fisheries of England*, London, 1880.
Hornell, J.,	*The Fishing Luggers of Hastings*, 'Mariners Mirror', 1938.
Innis, Prof H. A.,	*The Cod Fisheries*, Toronto, 1940.
Jenkins, A. C.,	*The Silver Haul*, London, 1967.
Jenkins, J. T.,	*The Sea Fisheries*, London, 1920.
	A History of the Whale Fisheries, London, 1921.
	The Herring and the Herring Fishery, London, 1927.
Judd, Stan,	*Inshore Fishing*, London, 1971.
Kendall, C.,	*God's Hand in a Storm*, Hull, 1870.
Leatham, J.,	*Fisherfolk of the Northeast*, Turriff, 1937.
Leather, John,	*Northseamen*, Lavenham: Dalton, 1971.
Lubbock, Basil,	*The Arctic Whalers*, Glasgow, 1937.
Lumley, Adrian, (and others)	*The Handling and Stowage of White Fish at Sea*, London, 1929.
Lund, Paul and Ludlam, Harry,	*Trawlers Go To War*, Slough, 1971.
March, Edgar J.,	*Sailing Drifters*, London, 1952.
	Sailing Trawlers, London, 1953.
	Inshore Craft of Great Britain, Newton Abbot, 1970.
Marwick, W. H.,	*Scotland in Modern Times*, London, 1964.
Mather, E. J.,	*Nor'ard of the Dogger*, London, 1887.
Matheson, Colin,	*Wales and the Sea Fisheries*, Cardiff, 1929.
Mitford, William,	*Lovely She Goes*, London, 1969.
Moore and Moore,	*The History and Law of Fisheries*, London, 1903.
Murie, Dr James,	*Report on the Sea Fisheries and Fishing Industries of the Thames Estuary*, London, 1903.
Nall, J. G.,	*Great Yarmouth and Lowestoft*, London, 1867.
Noall, Cyril,	*Cornish Seines and Seiners*, Truro: 1972.
Ogilvy, James S.,	*Relics and Memorials of London City*, London, 1910.
Ommanney, F. D.,	*A Draught of Fishes*, London, 1965.
Pain, E. C.,	*The Last of Our Luggers and the Men Who Sailed Them*, Deal and Sandwich, London, 1929.

Patterson, A. H.,   *Notes of an East Coast Naturalist*, London, 1904.

Phillips, Hugh,   *The Thames About 1750*, London, 1951.

Philpots, John R.,   *Oysters and All About Them*, London, 1890.

Piper, Steven,   *The North Ships*, Newton Abbot, 1974.

Pulling, Alexander,   *The Laws, Customs, Usages and Regulations of the City and Port of London*, London, 1854.

Radcliffe, W.,   *Fishing From Earliest Times*, London, 1926.

Rae, B. B.,   *The Lemon Sole*, London, 1965.

Reynolds, Stephen,   *Alongshore*, London, 1910.

     A Poor Man's House, London and New York, 1909.

Richardson, W.,   *The Practical Fishmonger and Fruiterer*, 4 vols., London, 1914–17.

Robinson, James,   *The Whole Art of Curing, Pickling and Smoking Meat and Fish*, London, 1847.

Samuel, A. M.,   *The Herring: Its Effect on the History of Britain*, London, 1918.

Scoresby, William,   *Account of the Arctic Regions*, 2 vols., Edinburgh, 1820.

Shaw, George,   *Our Filey Fishermen*, London, 1867.

Singer, Burns,   *Living Silver*, London, 1957.

Smith, Edwin Green,   *The Sea Fishing Log of Edwin Green Smith*, Grimsby, 1884–8.

Smith, Simon,   *The Herring-Busse Trade*, London, 1641.

Smith, W. C.,   *A Short History of the Irish Sea Fisheries During the Eighteenth and Nineteenth Centuries*, Liverpool and London, 1923.

Street, Philip,   *Beyond the Tides*, London, 1955.

Swinburne, T. A.,   *The Rig of Fishing Boats on the East Coast of Scotland*, Edinburgh, 1883.

Taylor, Benjamin,   *Storyology – Essays of Folklore and Sealore*, London, 1900.

Tressler, D. K. and Lemon, J. McW.,   *Marine Products of Commerce*, New York, 1957.

Tunstall, Jeremy,   *The Fishermen*, London, 1962.

'Viking',   *The Art of Fishcuring*, Aberdeen, 1911.

Villiers, Alan,   *The Deep Sea Fisherman*, London, 1970.

Walker, Dora M.,   *Freemen of the Sea*, London, 1951.

Wheeler, W. H.,   *The North Sea*, London, 1908.

White, P.,   *Observations on the Present State of the Scotch Fisheries*, Edinburgh, 1791.

Whymper, F.,   *Fisheries of the World*, London, 1883.

Willis, Jerome,   *Trawlermen's Town*, London, 1940.

Wilson, Gloria,   *Scottish Fishing Craft*, London, 1965.

	More Scottish Fishing Craft, London, 1968.
Wimpenny, R. S.,	*The Plaice*, London, 1953.
Wood, Walter,	*North Sea Fishers and Fighters*, London, 1911.
	Fishing Boats and Barges from the Thames to Land's End, London, 1922.
	Fishermen in War Time, London and Edinburgh, 1918.

PICTURE CREDITS

The author gratefully acknowledges the following for permission to reproduce illustrations:

Peter Brady: P. 28 (bottom)
British Library Board: P. 39 (top and bottom)
British Trawler Federation: P. 295 (bottom)
Fishing News: Pp. 86 (bottom), 298, 307 (top and bottom)
Great Yarmouth News Agency: Pp. 63, 241 (top and bottom)
Ford Jenkins: Pp. 79, 147 (bottom), 148 (top), 155, 253, 276–277, 279 (top), 280–281
City of Kingston-upon-Hull, Museums and Art Galleries: Pp. 247, 250–251, 266
National Gallery of Scotland: P. 52
Country Life Archive, National Museum, Edinburgh: Pp. 189, 190 (bottom)
Norfolk Museum Service, Great Yarmouth Museums: Pp. 70 (top), 147 (top)
Iain Rae: Pp. 133, 151, 177, 181, 201, 221, 278, 297
Royal Nation Mission to Deep Sea Fishermen: Pp. 13, 15, 33, 50–51, 59, 68, 70 (bottom), 80, 83, 85, 89, 92, 100 (all), 122, 125 (top and bottom), 128, 131, 134, 136, 137, 139 (all), 142 (top and bottom), 144, 148 (bottom), 158, 161, 165, 169, 192 (top), 203, 208, 210 (top and bottom), 230, 237, 243, 246, 259 (top and bottom), 262, 271, 279 (bottom), 282 (bottom), 302 (bottom)
Scarborough Library: P. 153 (top)
National Maritime Museum, London: Pp. 163, 215 (top and bottom), 216, 219, 225 (top and bottom), 232, 237 (bottom)
National Maritime Museum, London – Edgar March Collection: Pp. 95, 98–99, 126–127, 130
Scottish Fisheries Museum, Anstruther: Pp. 178–179, 190 (top), 192 (bottom), 262 (bottom)
Trawling Times: P. 308 (top)
Whitby Museum – F. M. Sutcliffe Collection: Pp. 19, 22–23, 28 (top), 152 (bottom), 197, 198–199, 200 (top and bottom), 203 (bottom)
White Fish Authority: P. 308 (bottom)
The drawings on pp. 72, 154, 250–251, 261 are by K. C. Lockwood.
The diagrams on pp. 272–273, 312 are by Martin Steenson.
The photographs on pp. 300–301, 305 are by the author.

INDEX

Aberdeen, 47, 180, 183, 240, 248, 258
Aberystwyth, 236
Adams, Robert, 53
Adriatic, 49, 289
Agriculture and Fisheries, Ministry of, 10
Allard, Peter, 10
Allithwaite, 235
Almanack des Gourmands, 242
Amble, Northumberland, 299
Anglesey, 236
Anson, Peter, 9
Anstruther, Fife, 10, 175, 180, 185
Arbroath, 183
Ardrishaig, 175
Ardrossan, 183
Auctioning of catches, 84, 86
Ayr, 183

Bacon, Admiral Sir Reginald, 290, 291
Bagshawe, J. R., 193
Baker, Robert, 164
Baldies, 176
Baltic, 47, 49, 77, 207, 213
Banff, 58, 175, 183
Barges, 213
Barking, 37, 66, 71, 73, 74, 77, 78, 206, 248
Barnard, Tom, 166
Barnstaple, 40
Barrow, 151
Bath, 74, 220
Baulks, 35
Bawleys, 207, 208, 212, 293
Bear Island, 254
Beccles, 44
Beer, Devon, 218
Beesands, Devon, 227
Benfleet, 36
Benkels, Willem, 46
Bergen, 40
Berwick, 57, 172, 240
Bethel ships, 128–144
Bideford, 40
Billingsgate, 21, 32, 66, 69, 71, 76, 81, 84, 85, 87, 96, 145, 206, 207, 208, 236, 249
Biscay, bay of, 254
Blackbond, Frederick, 164
Blackpool, 207, 235
Blackwater, 36

Blyth, Northumberland, 53
Board of Trade, 164, 167, 228, 263
Bogs, 217
Bolitho, T. & Sons, 232
Bones, Alexander, 37
Boston, 38, 92, 248
Botany Gut, 111
Bournemouth, 218
Boys' Own Paper, 135
Brancaster, 37
Brand, Osmond, 117
Breydon Water, 44
Brightlingsea, 207, 213
Brighton, 83, 214
Bristol, 38, 40, 53, 74, 220, 226, 249
British Society (fishing company), 57
Brittany, 211, 231
Brixham, 73, 74, 75, 80, 82, 92, 118, 218, 224, 236, 249, 254, 292
Broads, the, 44, 149
Brunel, I. K., 223
Buchan, Mr. & Mrs. John, 10
Buchanan, Rev. John, 55
Buckhaven, 58
Buckie, 180, 182, 183
Burghead, 58, 182
Burnmouth, 174, 192
Burntisland, 58
Busses, 41, 47, 53, 54, 55, 56

Cabot, John & Sebastian, 40
Caerarfon, 236
Campbeltown, 183
Canoes, dug out, 38
Capstick, Dick, 10
Cardigan Bay, 35, 236
Carew, Richard, 229
Carson, Sir E., 289
Cawsand, 232
Cecil, Lord, 41
Cawsand, 232
Cecil, Lord, 41
Cellardyke, 58
Channel Islands, 212
Chichester, 209
China, 78
Church of England Temperance Society, 134, 135
Cinque Ports, 44, 45, 62
Clacton, 36
Clarkson, Gordon, 10

Clothes, fishermen's, 17, 18, 29, 92
Clyde, 58
Coffee Smith's fleet, 92
Colchester, 36, 37
Coldingham, 174, 182
Coleridge, Samuel Taylor, 65
Colne, 36, 211
Columbia fleet, 92
Copers, 121–124, 126–127, 131–133, 143
Cornwall, 83, 214, 220, 223, 229, 234
Countryman, the, 194
Cove, 182
Cowes, 218
Cran Measurement Act, 59
Crisp, Thomas, R.N.R., 289
Cromarty, 182
Cromer, 38
Crooks, J., 288
Cullen, 58
Curing, 59
Cutters, 81, 119, 120, 121
Cyprus, 181

Dagons, 234
Dalrymple, Alexander, 78
Dann, William, 164
Dardanelles, 289
Dart, river, 35
Dartmouth, 231
Deep Sea Fishermen, Royal National
    Mission to, 9, 131–144, 159
Defoe, Daniel, 35, 46, 61, 218, 231
Dempster, George, 78
Denmark, 43, 47, 49, 306
Denovan, J. F., 57
Dickens, Charles, 61
Distinguished Service Cross, 290
Dogger Bank, 26, 38, 40, 75, 91, 97,
    106, 108, 111, 113, 124, 129, 159,
    273, 303
Dover, 44, 45, 74, 218, 249
Drifters (drivers), see also under
    luggers, 78, 145–146, 191, 224, 292
Dublin Bay, 226
Dunbar, 58, 171, 185
Dunbeath, 172, 173, 182
Dunwich, 38
Durham, 196, 205

Edinburgh, 84, 172, 185, 240
Eddystone lighthouse, 138
Eelboats, 87
Egypt, 181
Ellis, Jack, 10, 261
Emsworth, 211
Encouragement of White Herring
    Fishery Act, 56
Erith, 37
Estonia, 181
Evans, Captain, 212

Exe, 209, 211
Exeter, 40, 74
Eyemouth, 58, 174, 182, 240, 242

Fal river, 209
Falmouth, 211, 222
Faroe Islands, 40, 69, 106, 254
Fife, 87, 172
Fifies, 176, 276
Filey, 204
Findon, 240, 242
Finland, 181
Fisheries Board, Scottish, 57, 58, 272
Fisheries Regulation Act, 1888, 273
Fisherlads' Institute, 116
Fisherrow, 185
Fishmongers' Company, 10, 85
Flag signals, 103
Flamborough, 71, 123, 124, 264
Flanders, 44, 49, 53
Fleetwood, 235, 236, 248, 249, 254,
    261, 287, 292, 299
Flookburgh, 235
Folkestone, 82, 218
Forth, Firth of, 57, 58, 172, 173, 297
Foulness, 208
Fowey, 231
France, 44, 45, 46, 47, 49, 53, 54, 73
Fraserburgh, 58, 174, 180, 182, 183
Fyne, loch, 58, 175

Gaffers, 213
Gamecock fleet, 92, 265
Gardner, Gerald, 10
Garths, 35
Gawne, John, 156
Gentleman, Tobias, 41, 46
Germany, 82, 181
Glasgow, 238
Goodchild, Skipper, 132
Goodwin Sands, 218
Gorleston, 77, 91, 108, 135, 186, 248
Goryds, 35
Grafton, Duchess of, 137
Graham, Dr. Michael, 284, 293
Granton, 183
Gravelines, 212
Gravesend, 69, 207
Great Grimsby Ice Company, 133, 248
Great Grimsby Steam Trawling Com-
    pany, 246
Great Northern Fleet, 92, 133
Great Yarmouth, 10, 41, 44, 45, 47, 49,
    53, 54, 61, 62, 64, 71, 77, 78, 91, 106,
    110, 118, 129, 135, 145, 157, 166,
    185, 186, 188, 192, 193, 238, 239,
    240, 249, 254, 284, 294
Greece, 41, 181
Green Smith, Edwin, 71
Greenland, 254, 310

Greenock, 183
Greenwich, 10, 37
Grenfell, Dr Wilfred, 120, 138
Grimsby, 27, 66, 67, 69, 71, 76, 77, 78,
   81, 82, 84, 91, 105, 109, 114, 115, 116,
   159, 162, 164, 193, 204, 218, 219, 240,
   245, 246–249, 252, 254, 260, 269, 270,
   272, 280, 286, 287, 293, 299, 304
Grog-ships, see under Copers
Guorts, 35
Gurgites, 35

Haberson, Ada, 94
Hague, the, 306
Hallsands, Devon, 217, 227
Hamble, 209, 218
Hamburg, 56
Hamling, Thomas & Company, 10
Hanseatic League, 27, 43, 57
Hardy, Sir Alister, 272
Hartlepool, 124, 205
Harwich, 66
Hastings, 44, 214, 217, 218
Hatch boats, 69
Hebrides, 55
Heligoland, 91
Hellyer, Robert, 80, 81
Helmsdale, 57, 58, 182, 183
Herring Industry Board, 10, 176
Hewett, John, 108, 236
   Samuel, 66, 67, 77, 78, 80, 91
   Scrymgeour, 67
Hill, David Octavius, 52, 53
Hog-boats (hoggies), 214
Holborn Guardians, 116
   66, 67, 204, 207, 209, 274
Holland, 27, 41, 44, 46, 49, 53, 54, 62,
Holy Loch, 209
Holyhead, 236
Hopeman, 182
Howe, Harry, 288
Hoylake, 236, 254
Hughes, Peter, 117
Hull, 10, 27, 29, 66, 76, 78, 80, 81, 91,
   115, 118, 121, 123, 159, 162, 240,
   248, 249, 254, 257, 265, 267, 286, 287,
   293, 294, 299, 304
Humber, 61, 76, 81, 84, 111, 162, 196,
   205, 207, 246
Hunters, William, 202
Huxley, Professor T. E., 27, 175, 269,
   270
Hythe, 44

Iceland, 38, 40, 49, 69, 106, 254, 255,
   264, 294, 303, 306, 309, 310
Ilfracombe, 220
Insurance, 110
International Fisheries Exhibition,
   1883, 269

International Council for the Explora-
   tion of the Sea (I.C.E.S.), 272, 293
Inverness, 183, 240
Ipswich, 36, 211
Italy, 41, 47, 229

Jagers, 47
Jan Mayen Island, 245
Jarrow, 245
Jellicoe, Lord, 289
Jenkins, Tom, 227
Jenkinson, Matthew, 202
Johnson, Dr, 209

Keddles, 35
Keppers, 35
Kettle nets, 35
Kingfisher, Charles, 10
Kings Lynn, 38
Kirkaldy, 183
Kirkwall, 183
Knox, Ken, 10

Labrador, 26, 40, 138, 254
Largo Bay, 213
Latvia, 181
Laws, fishing, 54, 55, 58, 59
Leather, John, 9, 212
Leeds, 114
Legends, 15, 191
Leicester, 115
Leigh, 207, 208
Leisure Hour Magazine, 239
Leith, 53, 183, 303
Leleu's fleet, 92, 129, 143
Lerwick, 175, 183
Lewis, Charles, 10
Lewis, Jack, 10
Licences, fishing, 53
Lill, John, 118
Lincoln, 115
Lisbon, 211
Lithuania, 181
Liverpool, 69, 76, 83, 226, 235, 236
Lizard, the, 222, 224
Lockwood, K. C., 10
London Hospital, 96, 138
Longboat, Viking, 38
Lossiemouth, 58, 176, 182, 297
Lowestoft, 53, 62, 77, 78, 106, 111, 117,
   145, 149, 150, 153, 154, 157, 185,
   228, 236, 248, 249, 254, 272, 276,
   278, 283, 285, 294
Luggers, 62, 64, 146, 149, 150–157,
   217, 224
Lybster, 58, 182
Lyon Dean, Dr. W. J., 10, 176

Macduff, 58
Malta, 181

Man, Isle of, 176, 226, 234
Manchester, 83, 254
Mann, John, 160
March, Edgar, 9, 224, 245
Marine Biological Association, 272
Maritime Trust, 10
Markham, R. C., 10
Mather, Ebenezer, 128, 129, 135, 159
Medical remedies, 96
Mediterranean, 49, 69, 81
Medway, 36
Meldrum, A. N., 194
Menai straits, 35
Merchant Shipping Act, 114, 115, 263
Mevagissey, 232
Milford Haven, 138, 236, 248, 287, 292
Minch, the, 296
Minehead, 35
Minesweepers, 289
Mission ships, 128–144, 160, 265
Mission to Deep Sea Fishermen, Royal
    National, 9, 131–144, 159
Mitchell, A. T., 164
Montrose, 183
Mooney, George, 109
Moray Firth, 172, 180, 183, 242
Morecambe, 35, 204, 235
Morgans fleet, 83, 129, 143
Morocco, 236, 254
Mounts Bay, 174, 221, 222
Mules, 226
Mumbles, the, 209
Mumble bees, 226
Mundahl, Mr, 164
Mure, Robert, 10
Murmansk, 254
Murray, Miss Mary, 10, 185, 187, 188,
    193
Museums, Great Yarmouth, 10;
    Greenwich, 10;
    Hull, 10;
    Scottish Fisheries, 10
Musselburgh, 182, 185

Nabbies, 175
Nashe, Thomas, 62
National Lifeboat Institution, 174
National Maritime Museum, Greenwich,
    10
Naval Exhibition, Chelsea, 1891, 138
Navigation Act, Cromwell's, 54, 61
Newcastle, 238
Newfoundland, 29, 40, 41, 74, 138, 236,
    310
Newhaven (Edinburgh), 53, 182, 185
Newlyn, 94, 224, 227, 228, 232, 292
Newquay, 232
Nickeys, 234
Nobbies, 235
Norham Castle, 49

North East Atlantic Fisheries Con-
    vention (N.E.A.F.C.), 293
North Foreland, 217
North Sea Fishers and Fighters, 265
North Sea Fishing Act, 124
North Shields, 205, 248, 283
Norwich, 62

Orkney, 183
Ostende, 129
Otranto, straits of, 289
Overbury, Thomas, 118

Pacific, 49
Padstow, 138
Paget-Tomlinson, Edward, 10
Pall Mall Gazette, 132
Papper, William, 117
Patterson, A. H., 110
Payment of Wages Act, 115
Pegwell Bay, 218
Penzance, 222, 223
Perth, 183
Peterhead, 58, 172, 173, 180, 182, 183
Pickling, 47, 193, 278
Plymouth, 40, 214, 217, 220, 222, 227,
    248, 254, 272, 292
Poland, 47, 181
Polperro, 221
Poole, 40, 209, 218
Port Isaac, 220
Portmahomack, 175
Portsmouth, 213
Portsoy, 58
Portugal, 41, 47, 49, 211, 310
Pratt's Ferry, 37
Prestonpans, 83, 209
Primrose, Hon. B. F., 174
Prussia, 181
Purdy, William, 245, 246

"Q" ships, 289
Queensborough, 212

Rae, Iain, 10
Railway transport, 77, 82, 83, 84, 85
Ramsgate, 74, 75, 76, 91, 116, 162, 218,
    249, 254
Raw fag nets, 35
Red Cross fleet, 92, 132, 133
Reed's Guide to Skipper and Mate
    examinations, 263
Reynolds, Stanley, 223, 243
Robin Hood's Bay, 196
Rockall, 236, 254
Romney, 44
Rosehearty, 182
Rotterdam, 118
Rowhedge, 167, 213

Royal National Mission to Deep Sea Fishermen, 9
Royal Provident Fund for Sea Fishermen, 166
Rumania, 181
Russia, 47, 171, 181, 265, 310, 311
Rye, 35, 44, 214, 254
Ryecroft, Fred, 117

St Ives, 232
St Kilda, 270
Salcombe, 81
Saltash, 223
Sandwich, 44
Scarborough, 38, 53, 74, 75, 76, 77, 82, 92, 109, 153, 174, 202, 204, 205, 226, 254
Schuyts, 62, 82
Scilly Isles, 226
Scotland, fishing industry, 57, 74, 171
Scotsman, The, 171
Scott, Sir Walter, 183
Scottish Fisheries, Board, 57, 58, 174
Museum, 10
Seagulls, legends, 15
Seahouses, 239
Selsey, 219
Seining, 229–233, 296, 297
Seven Stones reef, 226
Shaw, Rev. George, 204
Sheffield, 114, 254
Shellfish Association of Great Britain, 10
Shetlands, 40, 41, 47, 54, 58, 174, 183, 185, 186
Shipwrecked Fishermen and Mariners' Royal Benevolent Society, 166
Short Blue fleet, 66, 77, 78, 81, 91, 108, 128, 143, 248
Shrimpers, 207
Silver pits, 74, 75, 81, 270
Sixerns (sexaerings), 175
Skaffs (skaffies), 175, 177
Skillingers, 213
Sluys, battle of, 45
Smacks, 22, 66, 67, 69, 71, 74, 78, 104, 213
Society of Arts, 57
Society of Free British Fishers, 55, 56
Solent, the, 218
Solway Firth, 212
Southampton, 40, 212
Southend, 211
Southport, 207, 235
Southwold, 56
Spain, 41, 46, 47, 49, 181, 209, 229
Spitsbergen, 26
Staithes, 196
Sterne, Garth, 10
Stonehaven, 240

Stornoway, 54, 183, 213, 294
"Stow" boat, 206, 212
Stranraer, 183
Sudds, William, 75, 76
Sutherland, John, 173
Swansea, 236, 248
Sweden, 43, 47
Sykes, J. H., 123

Tagus river, 211
Tamar river, 220, 223
Taylor, Charles, 118
Taylor, Peter, 173
Teign, 211
Telford, Thomas, 172
Tenby, 236
Terschelling, 213
Tevereux, Edward, 115
Texel Island, 91
Thames, 36, 37, 53, 61, 81, 206, 207, 209, 211, 212
Thames Church Mission Society, 129, 131
Thomson, Rev. Charles, 58
Tilbury, 36
Toilers of the Deep, 135, 136, 137, 186, 207, 265
Tollesbury, 36, 213
Torbay, 219, 226
Torquay, 220
Trading, 41
Trapping, 35
Trawling, 18, 21, 24, 37, 38, 73, 74, 77, 78, 81, 84, 92, 100, 104, 107–108
Treves, Sir Frederick, 93, 96, 140
Tunstall, Jeremy, 303
Turkey, 181
Tweed, river, 49, 196
Tyne, 246

United Kingdom Mutual Insturance Company, 10
United Nations Law of the Sea Conference, 309, 310
United States of America, 74, 83, 181

Vatersay, 185
Veal, George, 124
Venice, 49
Victoria Cross, 289
Viner, Alan, 10

Wash, the, 53, 61, 67, 209
Washington, Captain John, 173, 174, 193
Watson Hall, Jonathan, 10
Watson, Parry, 10
Watt, Joseph, R.N.R., 290
Wesley, John, 227
West Indies, 209, 235

West Mersea, 213
Weymouth, 40
Wheatfell, Edward, 117
Whitby, 18, 22, 29, 153, 201, 205, 226
White Fish Authority, 10, 209
Whitstable, 212
Wick, 50, 57, 58, 172, 173, 176, 182, 183, 193
Wigtown, 183
Willcocks, Colonel John, 10
Winchelsea, 44
Winchester, 211
Wivenhoe, 213

Wolf Rock lighthouse, 227
Wood, Walter, 156, 265
Wood, William, 176
Woodyer, John, 238, 239
Woolwich, 37
Wrath, Cape, 175

Yare, river, 157
Yarmouth, see under Great Yarmouth

Zulu (fishing boat), 10, 176, 177, 178–9, 180, 181

Key to end papers: not drawn to scale